Ozone Reactions with Organic Compounds

Ozone Reactions with Organic Compounds

A symposium sponsored by the Division of Petroleum Chemistry at the 161st Meeting of the American Chemical Society, Los Angeles, Calif., March 29-30, 1971.

Philip S. Bailey,

Symposium Chairman

ADVANCES IN CHEMISTRY SERIES **112**

AMERICAN CHEMICAL SOCIETY

WASHINGTON, D. C. 1972

ADCSAJ 112 1-129 (1972)

Library of Congress Catalog Card 72-88560

ISBN 8412-0152-8

PRINTED IN THE UNITED STATES OF AMERICA

Advances in Chemistry Series

Robert F. Gould, *Editor*

FOREWORD

ADVANCES IN CHEMISTRY SERIES was founded in 1949 by the American Chemical Society as an outlet for symposia and collections of data in special areas of topical interest that could not be accommodated in the Society's journals. It provides a medium for symposia that would otherwise be fragmented, their papers distributed among several journals or not published at all. Papers are refereed critically according to ACS editorial standards and receive the careful attention and processing characteristic of ACS publications. Papers published in ADVANCES IN CHEMISTRY SERIES are original contributions not published elsewhere in whole or major part and include reports of research as well as reviews since symposia may embrace both types of presentation.

CONTENTS

PREFACE

The nine chapters which comprise this volume are based on papers presented at the symposium on Ozone Chemistry—Reactions with Organic Compounds. This, to my knowledge, was the fifth symposium or conference which was devoted (or included sessions which were devoted) exclusively to the ozone chemistry of organic compounds. The first was the broadly based International Ozone Conference in Chicago in November 1956. This was followed by symposia or sessions at the September 1959 ACS meeting in Atlantic City, the December 1963 Southwest Regional ACS meeting in Houston (both of which were either arranged or coarranged by the undersigned), and the August 1967 International Oxidation Symposium in San Francisco. It is hoped that these meetings will continue to be held every three to four years.

Interest in ozone–organic chemistry remains strong, not only because of its use and potential use in industrial synthetic chemistry and in the space program, but, perhaps more importantly today, its role in air pollution and its potential utilization in water pollution problems. More and more, ozone appears to be the brightest hope of the future for purifying sewage and industrial waste water. This is at least partially the result of research during the last 20 years which has called attention to the fact that ozone is much more versatile in its reactions than was supposed when it was earlier thought of purely as a double-bond reagent.

Although Harries reported upon the versatility of ozone, this was almost forgotten until recently because of the emphasis placed by Harries and others upon the ozonolysis reaction. Ozone can be described as a resonance hybrid of the following canonical structures: $:\ddot{O}{=}\dot{O}_{+}{-}\ddot{\underset{..}{O}}^{-}: \longleftrightarrow :\ddot{\underset{..}{O}}^{-}{-}\dot{O}_{+}{=}\ddot{O}: \longleftrightarrow \ddot{\underset{..}{O}}^{+}{-}\ddot{\underset{..}{O}}{-}\ddot{\underset{..}{O}}^{-}: \longleftrightarrow :\ddot{\underset{..}{O}}^{-}{-}\ddot{\underset{..}{O}}{-}\ddot{\underset{..}{O}}^{+}$. As such, it should be able to function as a 1,3-dipole, an electrophile, or a nucleophile. In the ozonolysis of olefins, acetylenic compounds, and aromatic groupings ozone is thought to behave predominantly as a 1,3-dipole although it is possible that it sometimes behaves initially as an electrophile in these reactions. It is likely that it also behaves as a 1,3-dipole in its reactions with various carbon–hydrogen bonds, which are thought to involve insertion. An insertion reaction of some description is also thought to occur in the ozone–hydrosilane and the organomercurial ozonations reported in this volume.

Much of the versatility of ozone in its reactions is the result of its electrophilic character. As an electrophile it reacts with certain olefins to give epoxides or rearrangement products thereof, with certain aromatics and heterocyclics, and with nucleophiles such as amines, sulfides, phosphines, phosphites, and others, perhaps including ethers. Regarding the reactions of ozone with carbon π systems, evidence is presented here that the initial attraction between ozone and the π system involves the formation of π complexes, examples of which have been observed and characterized for the first time. Reactions of ozone with amines and certain other nucleophiles are often as fast or faster than those with carbon–carbon double bonds. This book includes one such study involving aromatic amines. The reactions with sp^3 systems (carbon– or silicon–hydrogen bonds, etc.) are much slower but can occur readily when more reactive systems are not present. Examples of ozone as a nucleophile are few and are not included. There is no evidence that ozone can react as a radical at ordinary ozonation temperatures (room temperature or below). It is not paramagnetic, and any initiation of radical reactions *via* ozonation is probably caused by decomposition of hydrotrioxides (insertion products) rather than by ozone itself. In addition to the reactions mentioned, ozone reacts with carbon–nitrogen, nitrogen–nitrogen, and other unsaturated groupings.

Although interest is increasing in the versatility of ozone in its reactions with systems other than carbon π systems, there is still much to be learned concerning the ozonolysis reaction and other reactions between ozone and sp^2 carbon–carbon systems. Thus, six of the nine papers in this volume deal with such reactions. Four of these deal primarily with mechanistic aspects of these reactions, while the other two have more to do with the utility of ozonolysis although they also discuss mechanistic ideas.

The revival of interest in ozone–organic chemistry during the past 20 years, which has led to the five symposia or conferences mentioned above, is generally credited to Prof. Rudolf Criegee, Emeritus Director of the Institute of Organic Chemistry of the University of Karlsruhe, Germany. His extraordinarily detailed and precise investigations into the mechanism of the ozonolysis reaction transformed this classical reaction from an art into a science, dispelling unwarranted fears concerning it (based on misconceptions regarding the nature of the peroxidic ozonolysis products), and pointing the way to a host of additional theoretical and practical studies. Because of this pioneering work, Prof. Criegee can justly be called the father of modern ozone–organic chemistry.

During the past seven or eight years some excellent and novel studies have cast doubt upon the Criegee mechanism as the sole mechanism of ozonolysis. All of the newer mechanisms proposed, however, have acknowledged that the Criegee mechanism plays an essential, though perhaps sometimes minor, role in ozonolysis. Thus, the genius of Criegee is unblemished. In the view of this writer, time is showing that the Criegee zwitterion mechanism, with necessary refinements to account for recent stereochemical observations, is the major route to the various peroxidic ozonolysis products. The other mechanisms may compete and even become predominant under specialized conditions but are minor or nonexistent under normal ozonolysis conditions.

I agreed to arrange this symposium only if Prof. Criegee could be brought over as an honored participant. This was arranged through the help of the Petroleum Research Fund of the American Chemical Society, for which I am grateful. Because of the overwhelming contribution of Prof. Criegee to modern ozone–organic chemistry and of his great influence on my own professional life, especially during my 1953–54 stay in Karlsruhe, I am pleased to dedicate this volume to him, a true gentleman, scientist, and personal friend.

PHILIP S. BAILEY

Austin, Tex.
March 1972

1

Complexes and Radicals Produced during Ozonation of Olefins

PHILIP S. BAILEY, JAMES W. WARD, REX E. HORNISH, and FRED E. POTTS, III

The University of Texas at Austin, Austin, Tex. 78712

Ozonation of certain hindered olefins at −78°C or higher is known to give epoxides or rearrangement products. Some new examples are given. It has been proposed that a pi complex is formed initially during ozonation of olefins and that this either proceeds into 1,3-dipolar cycloaddition and ozonolysis products or forms a sigma complex which loses oxygen to give an epoxide. The greater the steric hindrance, the more that epoxide formation is favored. Chemical, NMR, and visible spectral evidence are presented for the formation of complexes between ozone on the one hand and 1-mesityl-1-phenylethylene, 1,2,2-trimesityl-1-methoxyethylene and similar compounds on the other, at −150°C. Ozonation of 1,2,2-trimesitylvinyl alcohol at temperatures ranging from −20° to −150°C gives the corresponding oxy radical. An ozone–olefin complex precursor to this is suggested.

In previous work it has been shown that a competition exists during ozonation of olefins between ozonolysis and epoxide formation (*1*). As steric hindrance increases around the double bond, the yield of epoxide or subsequent rearrangement products increases. This is illustrated with both old (*1*) and new examples in Table I for purely aliphatic olefins and in Table II for aryl substituted ethylenes. It was suggested that the initial attack of ozone on an olefinic double bond involves π (pi) complex formation for which there were two fates: (a) entrance into 1,3-dipolar cycloaddition (to a 1,2,3-trioxolane adduct), resulting in ozonolysis products; (b) conversion to a σ (sigma) complex followed by loss of molecular oxygen and epoxide formation (Scheme 1). As the bulk

Table I. Percentage Epoxide from Ozonation of Aliphatic Olefins

Olefins	*Percentage*[a] *Epoxide*	*Reference*
tert-Bu-C=CH$_2$ CH$_3$ (on C)	10[b]	(*1*)
(*tert*-Bu)$_2$C=CH$_2$	15	(*1*)
(neopent)$_2$C=CH$_2$	29	present
(*tert*-Bu)$_2$C——C=CH$_2$ CH$_3$ CH$_3$ (on each C)	81	present
(neopent)$_2$C=CH-*tert*-Bu	91	present

[a] Yield of epoxide based on unrecovered olefin, usually determined by VPC.
[b] Yield of molecular oxygen accompanying epoxide formation.

Table II. Percentage Epoxide from Ozonation of Aryl Substituted Ethylenes

Olefins	*Percentage Epoxide*[a]	*Reference*
(C$_6$H$_5$)$_2$C=CH$_2$	15[b]	(*1*)
C$_6$H$_5$C=CH$_2$ *o*-tolyl (on C)	30[b]	(*1*)
C$_6$H$_5$C=CH$_2$ *o*-carboxyphenyl (on C)	40[c]	(*1*)
C$_6$H$_5$C=CH$_2$ Mesityl (on C)	82[d]	(*1*)
(Mesityl)$_2$C=CH$_2$	95[c]	present
(Mesityl)$_2$C=C(Mesityl)$_2$	[e]	present

[a] Yield of epoxide or rearrangement product(s) thereof based on unrecovered olefin.
[b] Molecular oxygen yield accompanying epoxide formation.
[c] Isolated yield of epoxide rearrangement product(s).
[d] VPC yield of epoxide rearrangement products.
[e] No products from attack on double bond found.

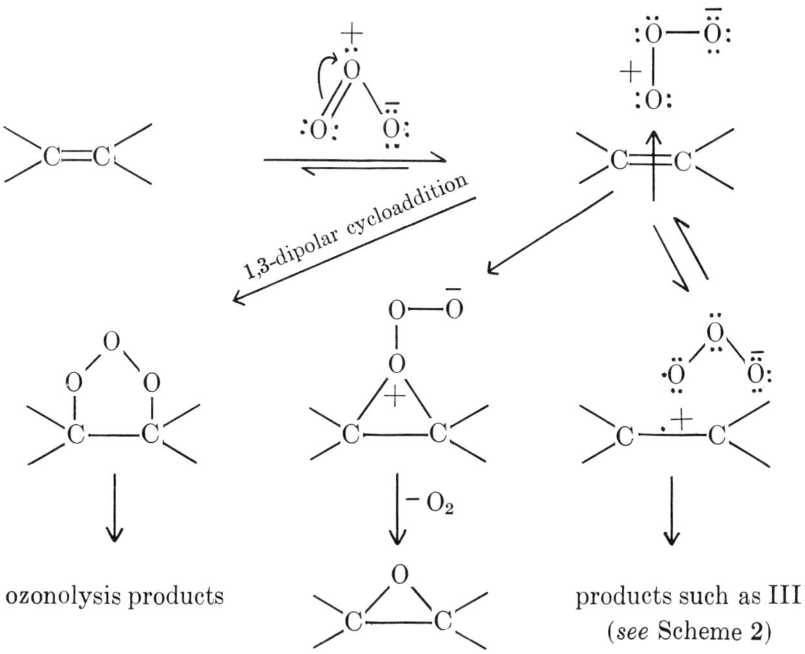

Scheme 1

of the groups around the double bond increases, the rate of 1,3-dipolar cycloaddition would be expected to decrease, and epoxide formation would be increasingly favored. In the case of tetramesitylethylene (Table II), steric hindrance is so great that there was no evidence for attack on the olefinic double bond to produce either ozonolysis products or an epoxide (or rearrangement products). It appeared that attack occurred only at the mesityl groups.

Additional evidence for the above picture of ozone–olefin interaction arose from ozonation of 1,2,2-trimesitylvinyl alcohol (I). Ozonation of a methylene chloride–methanol solution of the vinyl alcohol (I) with 1 mole equivalent of ozone at −78°C gave a yield of the corresponding oxy radical (III) of approximately 50% (2). Since ozone itself has no radical character, the most reasonable explanation appeared to be a third fate of a complex (*e.g.*, II, Scheme 2), which, because of excessive steric hindrance, undergoes neither 1,3-dipolar cycloaddition nor sigma complex formation but instead dissociates to a cation radical (IV) and the ozonate anion radical (V). The anion radical then abstracts a proton from IV to give III (2). This third fate is also illustrated in Scheme I. Supporting this mechanism are new data indicating that the yield of III is greater in polar than in nonpolar solvents (only 10–20% in pentane and 5–10% in carbon disulfide). This is reasonable on the basis that solvent polarity should increase the ease of dissociation of complex II (or some similar complex) through solvation of the resulting moieties. Another possibility is that proton abstraction occurs intramolecularly in complex II, after which dissociation to III and ·OOOH occurs.

To test these hypotheses further, 1-mesityl-1-phenylethylene (VIa) and 1,2,2-trimesitylvinyl alcohol (I) were ozonized at temperatures ranging from the usual −78°C (*1, 2*) to −150°C, and the acetate (VIIb) and methyl ether (VIIa) of the vinyl alcohol were ozonized at temperatures ranging from 0° to −120°C.

Ozonation of 1-mesityl-1-phenylethylene at −120°C gave the same results as ozonation at −78°C (*1*): evolution of molecular oxygen throughout the ozonation and the production of the corresponding epoxide or the subsequent rearrangement products. However, when the ozonation was carried out in Freon 12 at −150°C, ozone was rapidly absorbed, as at higher temperatures, but no molecular oxygen was evolved. As the ozone reacted, the reaction mixture changed from colorless to greenish to Burgundy red. No ozone passed into the iodide trap, and no precipitate formed in the reaction mixture. When the temperature of the reaction mixture was raised to about −140°C, the color disappeared, a gas was evolved, and a precipitate of the epoxide appeared.

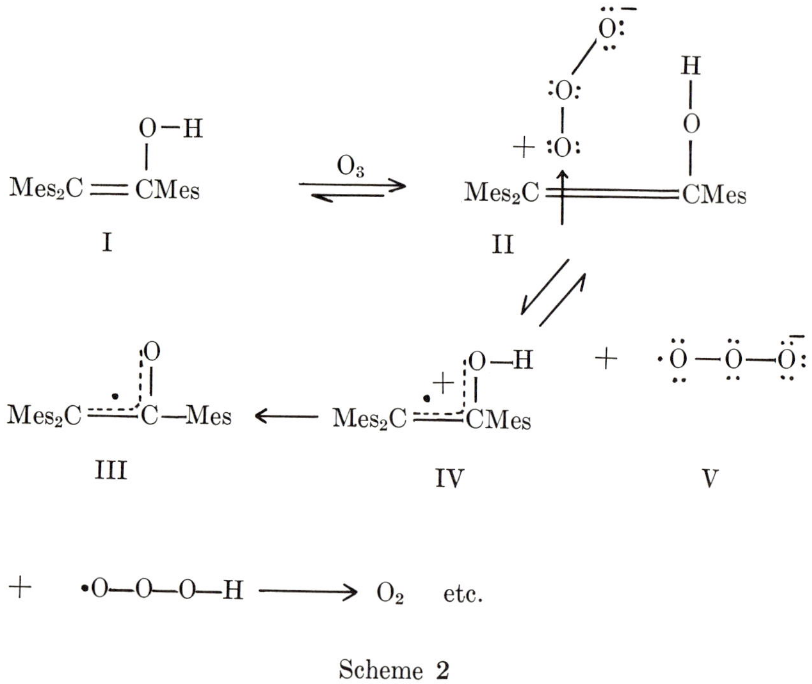

Scheme 2

Table III. NMR Absorption Peaks of 1-Mesityl-1-phenylethylene and Ozonation Products[a]

Olefin (−150°C)	*−150°C Complex*	*−135°C Reaction Mixture Containing Epoxide*	*Assignment*
7.69	7.52		phenyl protons
7.09	6.96	6.99	
6.74	6.58	6.68	mesityl ring protons
5.92	5.79		vinyl protons
5.03	4.90		
		2.97	former vinyl protons
2.29	2.15	2.17	mesityl methyl protons
2.06	1.97	2.04	

[a] The values shown are δ values expressed in ppm, relative to TMS.

In another experiment the gas evolved was shown to be molecular oxygen (Beckman oxygen analyzer) (*3*). Additional experiments showed that the evolved oxygen contained singlet oxygen, through its decoloration of rubrene and tetraphenylcyclopentadienone (*4*) and through mass spectral characterization of the rubrene–singlet oxygen product.

Further characterization of the ozone–mesitylphenylethylene complex produced at −150°C was done by NMR and visible spectral studies. The low temperature NMR spectra of the starting olefin, the red complex (ozonized olefin at −150°C) and the dilute reaction mixture at −135°C containing the epoxide of 1-mesityl-1-phenylethylene are described in Table III. The −150°C solutions of the olefin and the complex contain the same bands, the only difference being that the peaks were shifted slightly upfield in the formation of the complex. Such is typical of π complexes with very little charge transfer, such as iodine and tetracyanoethylene complexes of various aromatic molecules (*5, 6*). When the temperature of the ozonized reaction mixture was allowed to rise above about −145°C, the NMR spectrum changed, giving rise to the characteristic peaks of the epoxide of 1-mesityl-1-phenylethylene.

The two phenyl proton peaks in the olefin and the complex spectra must be conformational in origin since the olefin at room temperature has only one phenyl proton peak, at δ = 7.18 ppm. A LaPine molecular model of the molecule indicates that the phenyl group can rotate freely, but the mesityl group is firmly held out of the plane of the olefinic double bond. At room temperature there is only one phenyl proton peak as a result of free rotation, but at −150°C there are two as a result of the molecule's being "frozen" into one particular conformation.

Visible spectra of frozen solutions of ozone and of the ozone–mesitylphenylethylene complex in isopentane at −195°C (liquid N_2) showed clearly the production of a new band (at 450 mμ) in the formation of the complex. Only a trace of the characteristic (7) ozone band (at 575–625 mμ) was present in the spectrum of the complex.

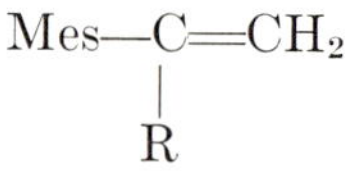

VI

(a) R = phenyl
(b) R = mesityl

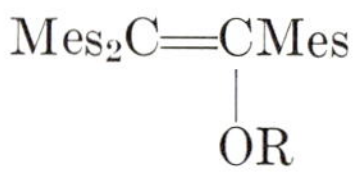

VII

(b) R = 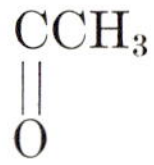

Ozonation of 1,1-dimesitylethylene (VIb) at −150°C also gave a color indicative of a complex, but because of the low solubility of the solvent, an NMR study was not possible.

Ozonation of 1,2,2-trimesitylvinyl alcohol (I.) at −150°C gave radical III strongly, just as at −78°C. Apparently, if a complex is formed first, the reaction over to the radical is so fast that the complex can not be detected.

Ozonation of the acetate (VIIb) and the methyl ether (VIIa) of vinyl alcohol I in Freon 11, Freon 12, carbon disulfide or methylene chloride at temperatures ranging from −50° to −145°C gave no ESR signal for a radical. Only in pentane or isopentane at −110°C or below was an ESR signal observed, and this proved to have been caused by the action of ozone on the solvent. In Freon 11 or 12 at −120°C, a greenish gray color indicative of a complex was obtained during ozonation of both the ether and the acetate. An NMR spectrum of the ozone–methyl ether (of trimesitylvinyl alcohol) complex was almost identical with that of the methyl ether itself at −120°C, with any apparent shifts being slightly upfield, as with the ozone–mesitylphenylethylene complex. The visible spectrum of the complex was similar to that of the complex of mesitylphenylethylene, except that the pure ozone absorption was slightly stronger.

The formation of these complexes is reversible, as one would expect. When a cold (−150°C) solution of the reactive olefin *cis*-3-hexene was added to the mesitylphenylethylene complex, the complex was destroyed with about one-fifth of the ozone reacting with the olefin to give *cis*- and *trans* 3-hexene ozonides and the other four-fifths producing the epoxide of mesitylphenylethylene, as determined by VPC and NMR spectral monitoring of the reaction. Similar treatment of the complex of 1-methoxy-1,2,2,-trimesitylethylene (VIIa) with *cis*-3-hexene at −120°C robbed the complex of all of the ozone, producing the 3-hexene ozonides.

Pi complexes have frequently been proposed in initial ozone–olefin interactions (*1, 8, 9, 10*). To our knowledge, however, the ones reported here are the first ever observed and characterized. The fact that complexes but no radicals were observed during ozonation of the ether and ester of trimesitylvinyl alcohol indicates that the hydroxy proton was essential to dissociation of the complexes to radicals. Further discussion of the significance of these complexes to ozonation of olefins, as well as a discussion of the π systems involved, will be given elsewhere.

Experimental

Materials. 3-*tert*-Butyl-2,3,4,4-tetramethyl-1-pentene and 1,1-dineopentyl-2-*tert*-butylethylene were prepared as described by Bartlett and Tidwell (*11*); 1,1-dimesitylethylene was prepared by the method of Snyder and Roeske (*12*); tetramesitylethylene was prepared by the method of Zimmerman and Paskovitch (*13*); trimesitylvinyl alcohol and its acetate were prepared by the methods of Fuson *et al.* (*14*). The methyl ether (VIIa) of the vinyl alcohol was prepared by using sodium hydride and methyl iodide in tetrahydrofuran; recrystallized from pentane, m.p. 161°–162°C. *Anal.* Calcd. for $C_{30}H_{36}O$: C, 87.33; H, 8.79. Found: C, 87.27; H, 8.70.

Ozonolyses. Ozonolyses and product determinations at ordinary temperatures were carried out as described in previous publications (*1, 2*).

Complex between Ozone and 1-Mesityl-1-phenylethylene. A small ozonolysis vessel containing 1.0–2.0 mmole of 1-mesityl-1-phenylethylene was purged with nitrogen and cooled to −150°C in an isopentane–liquid nitrogen bath, after which 1 ml of cyclopropane and 10–15 ml of Freon 12 were condensed in the vessel. The solution was purged with nitrogen, and an ozone–nitrogen stream (*3*) containing one equivalent of ozone was introduced into the reaction mixture. Some of the ozone condensed in the inlet tube but was slowly pushed into the reaction mixture by the nitrogen stream. As the ozone reacted, the color of the complex appeared and deepened. In some experiments *cis* and/or *trans*-3-hexene was added dropwise to the −150°C solution of the complex under a dry nitrogen atmosphere. Similar techniques were used with the other olefins which produced complexes.

NMR Spectra. The instrument used was a Varian Associates HA-100 NMR spectrometer equipped for taking spectra at low temperatures. The ozonation of 1-mesityl-1-phenylethylene was carried out as described above, except in the NMR tube itself using a syringe needle as the inlet. In some of the solutions studied cyclopropane was used as the standard. However, in all cases the δ values are relative to TMS.

Visible Spectra. These were taken on frozen isopentane solutions in a rectangular quartz cell cooled in liquid nitrogen in a quartz Dewar flask, using a Cary ultraviolet spectrophotometer.

ESR Spectra. These were recorded with a Varian Associates V-4502 spectrometer provided by NSF Grant GP-2090 to the Department of Chemistry. The techniques used are described in earlier papers (*2, 15*).

Singlet Oxygen Determinations. These were carried out as described by Murray *et al.* (*4*).

Acknowledgment

The support of this work by the Robert A. Welch Foundation (Grant F-042) and the Petroleum Research Fund of the American Chemical Society is gratefully acknowledged. The authors also thank Ben A. Shoulders for his patient help with the low temperature NMR determinations and Clifford Becker for the use of his low temperature quartz Dewar flask for visible and ultraviolet spectra.

Literature Cited

1. Bailey, P. S., Lane, A. G., *J. Amer. Chem. Soc.* (1967) **89,** 4473.
2. Bailey, P. S., Potts, F. E., III, Ward, J. W., *J. Amer. Chem. Soc.* (1970) **92,** 230.
3. Bailey, P. S., Reader, A. M., *Chem. Ind.* (1961) 1063.
4. Murray, R. W., Lumma, W. C., Jr., Lin, J. W. P., *J. Amer. Chem. Soc.* (1970) **92,** 3205.
5. Foster, R., "Organic Charge—Transfer Complexes," Chaps. 1–4, Academic, New York, 1969.
6. Foster, R., Fyfe, C., "Progress in Nuclear Magnetic Resonance Spectroscopy," Vol. 4, J. W. Emsley, J. Feenex, L. H. Sutcliffe, Eds., Chap. 1, Pergamon, New York.
7. Kirshenbaum, A. D., Streng, A. G., *J. Chem. Phys.* (1961) **35,** 1440.
8. Bailey, P. S., *Chem. Rev.* (1958) **58,** 925.
9. Murray, R. W., Youssefyeh, R. D., Story, P. R., *J. Amer. Chem. Soc.* (1967) **89,** 2429.
10. Williamson, D. G., Cvetanovic, R. J., *J. Amer. Chem. Soc.* (1968) **90,** 4248.
11. Bartlett, P. D., Tidwell, T. T., *J. Amer. Chem. Soc.* (1968) **90,** 4421.
12. Snyder, H. R., Roeske, R. W., *J. Amer. Chem. Soc.* (1952) **74,** 5820.
13. Zimmerman, H. E., Paskovitch, D. H., *J. Amer. Chem. Soc.* (1964) **86,** 2149.
14. Fuson, R. C., Chadwick, D. H., Ward, M. L., *J. Amer. Chem. Soc.* (1946) **68,** 389.
15. Bailey, P. S., Keller, J. E., Carter, T. P., Jr., *J. Org. Chem.* (1970) **35,** 2777.

RECEIVED May 20, 1971.

2

Ozonolysis: Formation of Cross Diperoxides

ROBERT W. MURRAY, JOHN W.-P. LIN, and DANIEL A. GRUMKE

University of Missouri, St. Louis, Mo. 63121

Parent and cross diperoxides are produced when tetra-substituted olefins containing suitable substituents are ozonized. Cross diperoxides are also produced when pairs of tetra-substituted olefins are ozonized together. Comparison samples of diperoxides are conveniently synthesized by treating the appropriate ketone with peracetic acid at low temperature. Peracetic acid oxidation of ketone pairs can also be used to prepare cross diperoxides. Low temperature NMR is used to study diperoxide stereochemistry as well as barriers to conformational isomerization.

The Criegee (*1*) mechanism of ozonolysis postulates that unsymmetrical olefins should give two zwitterions and two carbonyl compounds and hence postulates the possible formation of three different ozonides. This prediction has now been realized in a number of cases (*2–9*). It has also been shown that in many cases the ozonide stereoisomer ratio depends upon olefin stereochemistry in both normal (*3, 6–12*) and cross (*6–9*) ozonides. Since the original Criegee mechanism did not provide for these stereochemical results, a number of additional suggestions for the mechanism have been made (*6, 9, 13, 14*), all of which retain the fundamentals of the Criegee mechanism.

Criegee and Lohaus (*15*) and others have shown that tetra-substituted olefins do not usually give ozonide products but instead give dimeric and higher peroxides. Only in the special cases of negatively substituted (*18*) or cyclic olefins (*19*) do such compounds give ozonides. Presumably the failure to give ozonides is the result of the reduced reactivity of the ketonic carbonyl toward the zwitterion. If the ketone is present in large excess, such as when 2-pentene (*20*) or tetraphenylethylene (*21*) are ozonized in acetone solvent, then the corresponding ozonide is formed.

We have attempted to gain further information concerning the mechanism of ozonolysis by using conditions where ozonolysis might be

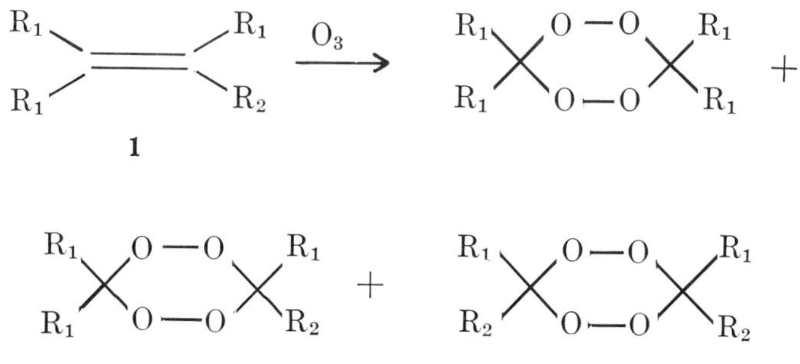

1

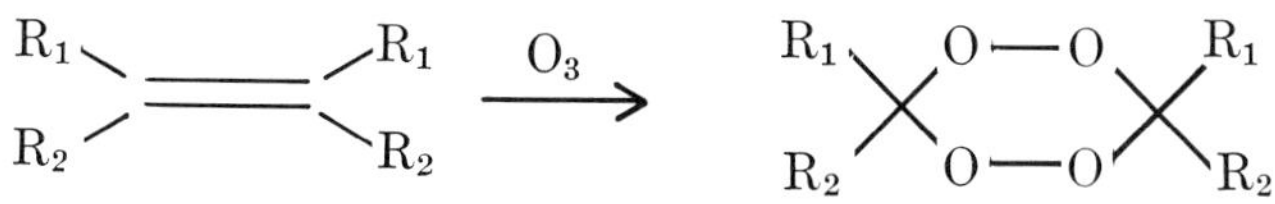

2

cis or trans

cis and/or trans

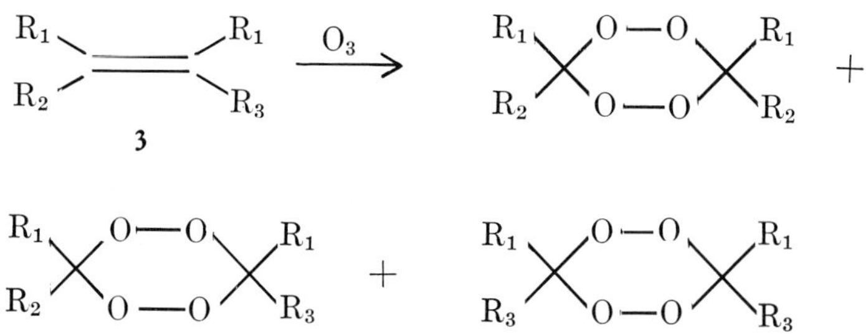

3

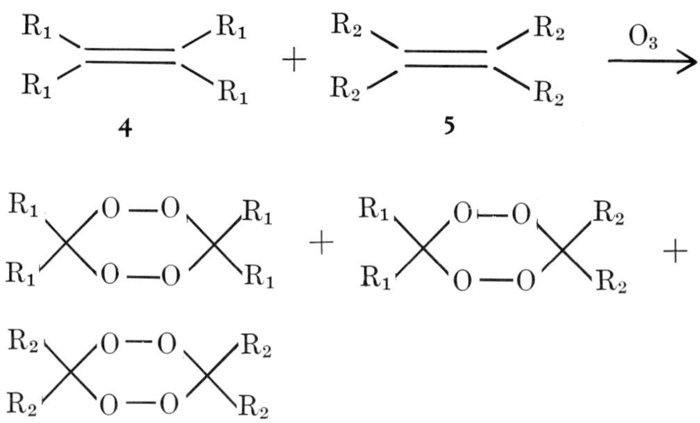

4

5

expected to give both normal and cross diperoxides. By choosing suitably substituted olefins we would ultimately want to examine diperoxide stereoisomer ratios and look for a dependence on olefin stereochemistry. In carrying out this approach there were at least four different variations to be considered. Ozonolysis of olefins type **1** could give both normal and cross diperoxides. Here one of the cross diperoxides is capable of existing as stereoisomers. Ozonolysis of stereoisomers of type **2** could give a single pair of stereoisomeric diperoxides. Ozonolysis of more complex tetrasubstituted olefins of type **3** could give both normal and cross diperoxides, each of which could exist in stereoisomeric forms. Finally, ozonolysis of pairs of symmetrically substituted olefins of types **4** and **5** could permit formation of both normal and cross diperoxides. In some of these cases more than one trans isomer of a diperoxide could be formed, depending upon axial-equatorial preferences of the substituents.

Experimental

Microanalyses were done by Schwarzkopf Microanalytical Laboratory, Inc., 56-19 37th Ave., Woodside, N. Y. 11377.

General Ozonolysis. A Welsbach T-23 ozonator was used as the ozone source. Its sample stream output was 0.5 mmole O_3/minute.

NMR Analyses. NMR spectra were recorded on a Hitachi-Perkin Elmer R-20 high resolution NMR spectrometer equipped with an R-202VT variable temperature accessory. Temperatures were determined by measuring peak separations in a methanol sample. Deuteriochloroform was used as solvent.

GPC Analyses. GPC analyses were carried out using a Varian-Aerograph Model A-700 gas chromatograph. Columns used were 3-10-ft columns containing 5–10% XF-1150 on chromosorb G.

Ozonolysis of Tetramethylene and *cis*-3,4-Dimethyl-3-hexene. A solution of tetramethylethylene (0.503 gram, 5.98 mmoles) and *cis*-3,4-dimethyl-3-hexene (0.692 gram, 6.17 mmoles) in 20 ml pentane was ozonized to 63% theoretical yield at −40°C. The product mixture was analyzed by GPC using an 8-ft 10% XF-1150 column at 85°C and a flow rate of 150 ml/minute. The mixture contained 25 mg acetone diperoxide, 36.7 mg 1,1,4-trimethyl-4-ethyl-2,3,5,6-tetraoxacyclohexane, and 23.2 mg methyl ethyl ketone diperoxide as determined by GPC using an internal standard. Total yield of diperoxides was 14%. The diperoxides were identified by comparing mp, infrared, NMR, and GPC data with those of authentic samples.

Ozonolysis of 2,3-Dimethyl-2-pentene. A solution of 2,3-dimethyl-2-pentene (1.04 grams, 10.6 mmoles) in 20 ml pentane was ozonized to 60% theoretical yield at −40°C. The product mixture was analyzed by GPC using an 8-ft, 10% XF-1150 column of temperature 85°C and flow rate 150 ml/minute. The mixture contained 0.503 mmole acetone diperoxide, 0.210 mmole 1,1,4-trimethyl-4-ethyl-2,3,5,6-tetraoxacyclohexane, and 0.12 mmole of methyl ethyl ketone diperoxide. Total yield of diperoxides was 31.1%. *Anal.* Calcd. for $C_7H_{14}O_4$: C, 51.84; H, 8.70. Found: C, 51.68; H, 8.51.

Ozonolysis of *cis*-3,4-Dimethyl-3-hexene. A solution of *cis*-3,4-dimethyl-3-hexene (98% pure, Chemical Samples Co.) (1.12 grams, 10 mmoles) in 50 ml pentane was ozonized at −62°C until the blue color of excess ozone was evident. A nitrogen stream was used to purge the solution of excess ozone. Pentane was then carefully distilled off at atmospheric pressure. A water aspirator (20 mm Hg) was then used to remove the ketone product. Treatment of this material with 10 ml of an 0.1*M* solution of 2,4-dinitrophenylhydrazine gave 2.33 grams of crude 2,4-dinitrophenylhydrazone. The crude product was recrystallized and identified as the 2,4-dinitrophenylhydrazone of methyl ethyl ketone, mp 115–116°C. Yield of the ketone was 92% based on olefin used.

The remainder of the ozonolysis mixture was distilled from 22°C (0.4 mm) to 55°C (0.03 mm) to give 0.549 gram of product. The crude product was distilled at 27°C (0.25 mm) to give 0.376 gram of a liquid. Analysis of this material by GPC showed that 70% of it was methyl ethyl ketone diperoxide (30% yield) identified by comparison with authentic material. The diperoxide had m.p. 7°–8°C. The remainder of the product was not identified.

Ozonolysis of *trans*-3,4-Dimethyl-3-hexene. A solution of *trans*-3,4-dimethyl-3-hexene (83% pure, Chemical Samples Co.) (1.325 grams, 10 mmoles) in 50 ml pentane was ozonized at −65°C until the blue color of excess ozone was evident. A nitrogen stream was used to purge the solution of excess ozone. Most of the pentane was carefully distilled off at atmospheric pressure, and methyl ethyl ketone was distilled off using the water aspirator. The ketone was identified through its 2,4-dinitrophenylhydrazone. The remaining reaction mixture was distilled at 25°C (2 mm) to give 0.396 gram of product. Analysis of this material by GPC showed it to be 95% pure (42.7% yield), and it was identified as methyl ethyl ketone diperoxide by comparing its infrared and NMR spectra and melting points with those of an authentic sample.

Ozonolysis of 2-Methyl-3-ethyl-2-pentene. A solution of 2-methyl-2-ethyl-2-pentene (99% pure, Chemical Samples Co.) (4.48 grams, 40 mmoles) in 60 ml of methylene chloride was ozonized at −65°C until excess ozone was present. After removal of excess ozone with nitrogen purging, methylene chloride and acetone were distilled off at atmospheric pressure. Diethyl ketone was removed under water aspirator pressure. The remaining reaction mixture was subject to short path distillation at 19°–23°C (0.05–0.02 mm) to give 2.774 grams of liquid (fraction A). Further distillation at 30°C (0.02 mm) gave 0.504 gram of liquid (fraction B). Analysis of fraction A by GPC showed that it contained acetone, diethyl ketone, and three diperoxides. The diperoxides were identified as acetone diperoxide, 1,1-dimethyl-4,4-diethyl-2,3,4,5-tetraoxocyclohexane, and diethyl ketone diperoxide by comparing infrared and NMR spectra and GPC retention times with those of authentic materials. GPC analysis of fraction B showed that it contained diethyl ketone diperoxide and another unidentified material. Yields as determined by GPC were: 4.26 mmoles acetone diperoxide, 5.19 mmoles 1,1-dimethyl-4,4-diethyl-2,3,5,6-tetraoxacyclohexane, and 3.34 mmoles of diethyl ketone diperoxide. Total yield of diperoxides was 32%.

Ozonolysis of Tetramethylethylene and Tetraphenylethylene. A solution of 0.259 gram (3.1 mmoles) of tetramethylethylene and 1.0 gram (3.1 mmoles) of tetraphenylethylene in 40 ml chloroform was ozonized to completion at room temperature. Chloroform was used as the solvent since tetraphenylethylene is fairly insoluble in pentane. Analysis of the reaction mixture by TLC and GPC showed only acetone diperoxide and benzophenone diperoxide to be present.

Preparation of Diethyl Ketone Diperoxide. To a solution of 0.37 mole peracetic acid (54 grams, 52% solution in glacial acetic acid) and 32 grams of 70% sulfuric acid we added, dropwise, 22 grams (0.25 mole) of diethyl ketone. The reaction mixture was kept at 2°–10°C for 80 minutes and then at 15°–17°C for 130 minutes. The organic layer was separated immediately and washed twice with 25-ml portion of 2*M* potassium bicarbonate solution. The total weight of the organic product was 21.48 grams. This material was distilled at 39°C (0.01 mm) to give 12.398 grams (46.2%) of diethyl ketone diperoxide. Redistillation of this material at 45°C (0.035 mm) gave pure diethyl ketone diperoxide, m.p. 8°–9°C. The room temperature NMR spectrum in $CDCl_3$ of the diperoxide had a multiplet at 1.8 (8H) and a triplet at 0.98 (12H, $J = 7.6$ Hz). *Anal.* Calcd. for $C_{10}H_{20}O_4$: C, 58.80; H, 9.86. Found: C, 58.62, H, 9.95.

Preparation of Methyl Ethyl Ketone Diperoxide. To a solution of 0.37 mole peracetic acid (54 grams, 52% solution in glacial acetic acid) we added, dropwise, 1.8 grams (0.25 mole) of methyl ethyl ketone. The reaction mixture was maintained at 8°–10°C during ketone addition. After addition was complete, the reaction mixture was stirred for 3.5 hours at 10°–18°C. Stirring was ended when the organic layer had reached maximum volume. The organic layer was separated and washed twice with 2*M* potassium bicarbonate solution. The crude diperoxide weighed 18.30 grams. The crude material was distilled at 44°–45°C (1.95 mm) to give 16.2 grams (73%) of pure methyl ethyl ketone diperoxide with mp 6.5°–8°C. The GPC analysis of the diperoxide on a 10-ft × ¼-inch XF-1150 column at 82°C and with a flow rate of 80 ml/minute, showing that it had a retention time of 48 minutes. *Anal.* Calcd. for $C_8H_{16}O_4$: C, 54.43; H, 9.14. Found: C, 54.39, H, 9.20.

Preparation of 1,1-Dimethyl-4,4-diethyl-2,3,4,5-tetraoxacyclohexane. To a cold mixture of 65.8 grams (0.45 mole) of 52% peracetic acid and 55 grams of 70% sulfuric acid we added, dropwise, over a one-hour period a mixture of 11.6 grams (0.2 mole) of acetone and 17.2 grams (0.2 mole) of diethyl ketone. During this addition the reaction mixture was maintained at 3°–6°C. The reaction mixture was then stirred vigorously at 6°–10°C for 3.5 hours. The solid reaction product was filtered off and washed with water. The organic layer of the filtrate was separated and washed twice with a 2*M* potassium bicarbonate solution. The total weight of the crude product obtained was 18.38 grams. Analysis of this material by GPC (5% XF-1150, 3-ft × ¼-inch, column temp −81°C) showed that the solid fraction contained 98.5% acetone diperoxide and 1.5% 1,1-dimethyl-4,4-diethyl-2,3,5,6-tetraoxacyclohexane. The liquid fraction (12.23 grams) contained 3.28% acetone diperoxide, 60.5% 1,1-dimethyl-4,4-diethyl-2,3,5,6-tetraoxacyclohexane and 31% diethyl ketone diperoxide. The pure 1,1-dimethyl-4,4-diethyl-2,3,5,6-tetraoxacyclohexane had m.p. 7°–8°C and b.p. 27.5°–30°C (0.02 mm). The NMR spectrum (room temperature, $CDCl_3$) of this material had a triplet at 0.975 (6H, $J = 7.6$ Hz), a singlet at 1.59, and a broad peak centered at 1.89. *Anal.* Calcd. for $C_8H_{16}O_4$: C, 54.53; H, 9.14. Found: C, 54.50; H, 9.03.

Preparation of 1,4,4-Trimethyl-1-ethyl-2,3,5,6-tetraoxacyclohexane. To a cooled solution of 5.11 grams (0.035 mole) of peracetic acid and 4.9 grams (0.035 mole) of 70% sulfuric acid we added, dropwise, with stirring 1.015 grams (0.0175 mole) of acetone and 1.34 grams (0.186 mole) of methyl ethyl ketone. The reaction mixture was maintained at 5°–10°C during this addition. After addition was completed, stirring at 10°C was continued for 3 hours. After addition of 20 ml of ice water the organic layer was separated and washed twice with 15-ml portion of 5% sodium bicarbonate. The organic layer was diluted to 50 ml with chloroform and analyzed by GPC using an 8-ft, 10% XF-1150 column at 85°C. The reaction mixture contained three diperoxides whose yields were determined by GPC using an internal standard. The products were acetone diperoxide (0.259 gram, 32.5%), 1,4,4-trimethyl-1-ethyl-2,3,5,6-tetraoxacyclohexane (0.41 gram, 47%), and methyl ethyl ketone diperoxide (0.16 gram, 20.5%). The room temperature NMR spectrum of 1,4,4-trimethyl-1-ethyl-2,3,5,6-tetraoxacyclohexane consisted of a triplet at 0.98 (3H, $J = 7.6$ Hz), and a broad singlet at 1.50 (11H). Its m.p. was 31°C.

Results and Discussion

One of our earlier attempts to form a cross diperoxide used a variation of olefin types **4** and **5**. In this case equimolar amounts of tetraphenylethylene, **6**, and tetramethylethylene, **7**, were ozonized together. While ultimately both acetone diperoxide and benzophenone diperoxide could be isolated from the reaction mixture, it became apparent that these olefins have such different reactivities toward ozone that the tetramethylethylene was selectively ozonized. Only after most of the tetramethylethylene had been ozonized was the tetraphenylethylene attacked. The opportunity for cross diperoxide formation in this case is thus minimal.

By choosing olefins of comparable reactivity towards ozone this difficulty can be avoided. When approximately equimolar amounts of **7** and *cis*-3,4-dimethyl-3-hexene, **8**, were ozonized together all three predicted diperoxides were obtained—*i.e.*, acetone diperoxide, **9**, the cross diperoxide, 1,1,4-trimethyl-4-ethyl-2,3,5,6-tetraoxacyclohexane, **10**, and methyl ethyl ketone diperoxide, **11**.

All of the diperoxides obtained in this study are capable of undergoing conformational isomerization of the type described earlier (*22*) for acetone diperoxide. By using variable temperature NMR studies the barriers to such isomerizations can be obtained. We have measured these barriers for all of the diperoxides obtained in this study and will report the results elsewhere.

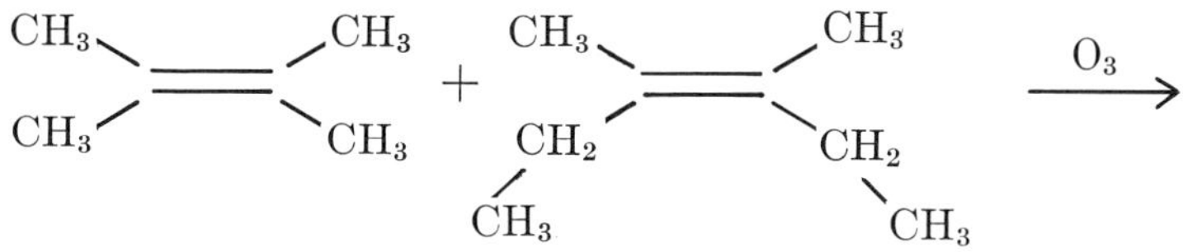

7 **8**

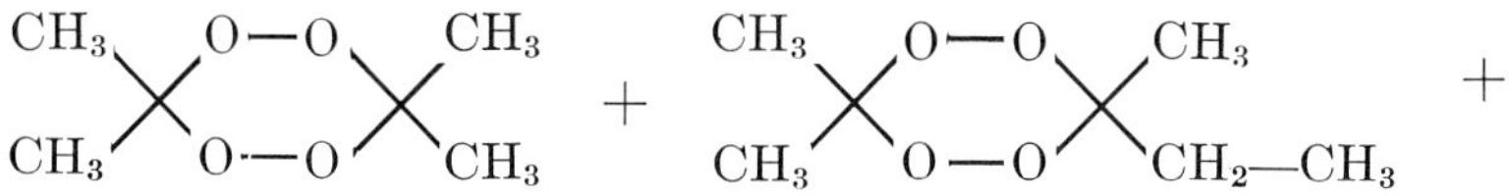

9 **10**

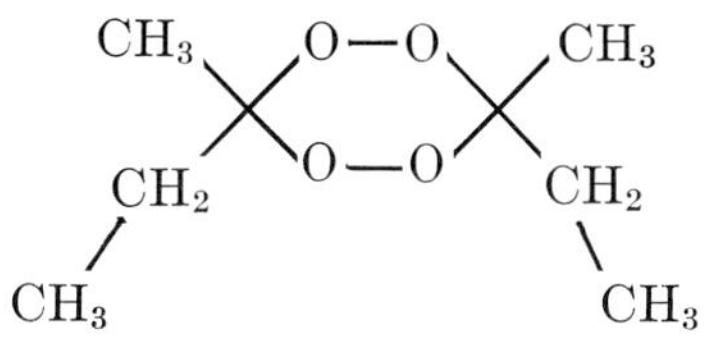

11

Low temperature NMR measurements are required in the present study, however, to attempt to determine the stereochemical makeup of diperoxides such as **11**. These results are described later in this paper.

The total yield of diperoxides obtained is usually low. Presumably most of the available zwitterions end up as polymeric peroxides. In the ozonolysis of **7** and **8**, for example, the total yield of diperoxides is 17.4%. The diperoxides were obtained in the ratio of 1:1.35:0.78 for **9:10:11** as compared with the statistically predicted ratio of 1:2:1.

We have found it useful to prepare authentic samples of the various diperoxides encountered by using a variation of the Baeyer-Villiger oxidation conditions. Oxidation of ketones at low temperatures using peracetic acid has been reported (*23*) to give diperoxides instead of the esters produced under Baeyer-Villiger conditions. Authentic samples of **10** and **11**, can be prepared, respectively, by the peracetic oxidation of acetone and methyl ethyl ketone jointly or methyl ethyl ketone alone. We are studying the mechanism of this interesting oxidation reaction.

An example of the ozonolysis of an olefin of type **1** is provided by the ozonolysis of 2,3-dimethyl-2-pentene, **12**. In this case the diperoxides **9**, **10**, and **11** are produced from a single olefin. A nomenclature consistent with that used for ozonides would describe **10** as the parent diperoxide and **9** and **11** as cross diperoxides.

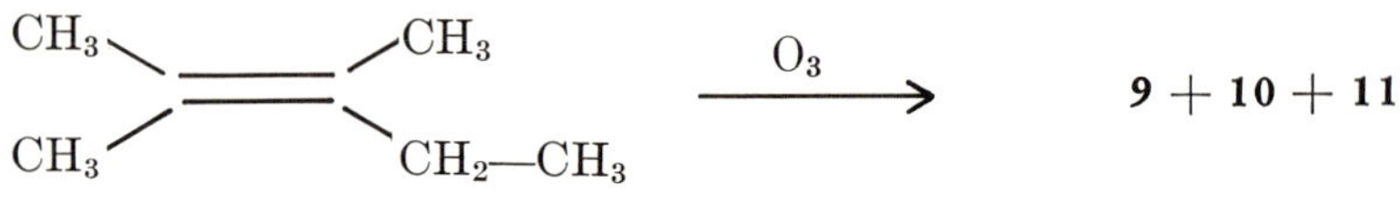

12

In this case the total yield of diperoxides was 27.6%. We are still trying to improve the yield of diperoxide products from these reactions; hence these yields should not be regarded as optimum values. At any rate, we have established that cross diperoxides can be formed so that we have an additional probe with which to examine the mechanism of ozonolysis problem.

It was now important to examine the question of a possible stereochemical influence on diperoxide formation. We have approached this problem initially by ozonizing olefins of type **2**. When either *cis*- or *trans*-3,4-dimethyl-3-hexene are ozonized, presumably a single stereoisomeric pair of diperoxides can be formed. In fact, this case is complicated by the possibility of two trans-diperoxide conformers being produced. The cis-diperoxide conformers are identical. Ozonolysis of *cis*-3,4-dimethyl-3-hexene, **8**, for example, could give the diperoxides, *cis*-1,3-dimethyl-1,3-diethyl-2,3,5,6-tetraoxacyclohexane, **11a**, and *trans*-1,3-dimethyl-1,3-diethyl-2,3,5,6-tetraoxacyclohexane, **11b**, with the latter capable of existing as conformers **11b** and **11b′** with trans-diaxial methyl and trans-diaxial ethyl substituents, respectively.

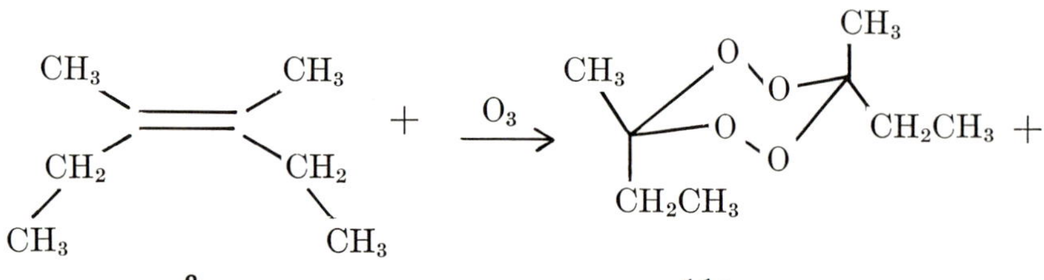

8 **11a**

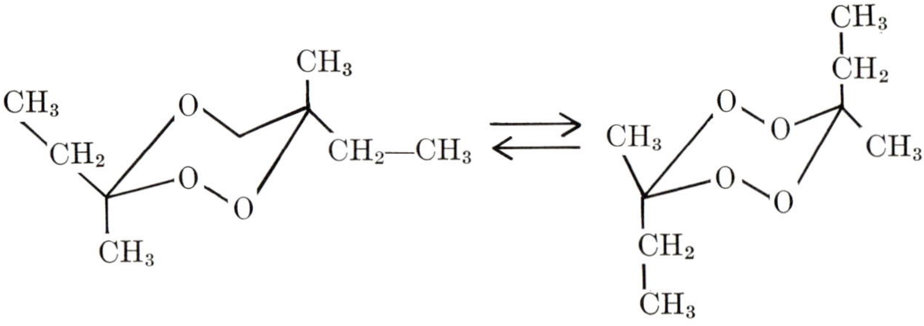

11b **11b′**

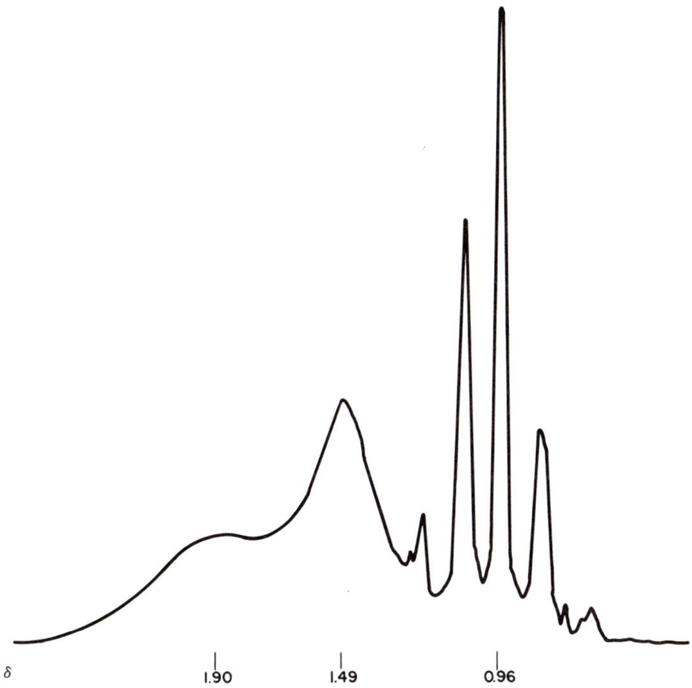

Figure 1. Room temperature NMR spectrum of methyl ethyl ketone diperoxide

Similar stereochemical possibilities are present when **11** is obtained from *trans*-1,3-dimethyl-3-hexene, **13**, as a cross diperoxide from the ozonolysis of **12** or from the peracetic acid oxidation of methyl ethyl ketone.

In fact, in all of these cases **11** appears as a single GPC peak under various analytical conditions. The room temperature NMR spectrum of **11** obtained from all of the sources described is also identical (Figure 1), consisting of a triplet at 0.96 (6H, $J = 7.6$ Hz), a broad singlet at 1.49 (6H), and a broad peak at 1.90 (4H) (chemical shift values given are δ values relative to internal TMS). The low temperature (*ca.* −40°C) NMR spectrum of **11** in each case is also identical and consists of (Figure 2) a triplet at 1.0 ($J = 7.6$ Hz), a singlet at 1.31, a singlet at 1.8, a quartet centered at 1.73 ($J = 7.6$ Hz), and a quartet centered at 2.28 ($J = 7.6$ Hz).

These NMR results can be interpreted as follows. At room temperature rapid conformational isomerization is occurring in **11**. This leads to a single broad peak at 1.49 for the axial and equatorial methyl groups and a single broad peak at 1.90 for the axial and equatorial methylene groups. The triplet at 0.96 is then assigned to the methyl protons of the ethyl groups which are not affected by the conformational isomerization. At low temperature the isomerization process is slowed to the point where the chemical shifts of the protons in the individual conformers can be detected. The triplet at 1.0 is still assigned to the methyl protons of the ethyl group. The two quartets at 1.73 and 2.28 are then assigned to the axial and equatorial methylene groups, respectively. The two singlets at 1.31 and 1.8 are assigned to the axial and equatorial methyl groups, respectively. The equatorial methylene and methyl groups are assigned the low field absorptions on the basis of ring current effects (*24*).

The axial and equatorial methyl and methylene absorptions have essentially equal peak areas. This spectrum is therefore consistent with the product's being the cis isomer, **11a**. For the trans isomer one might expect a preference for the conformer with the diequatorial ethyl groups, **11b**. In this case the axial and equatorial absorptions should not have equal intensities. Such a preference is expected to be small, however, and the NMR spectrum must be regarded as consistent with the trans

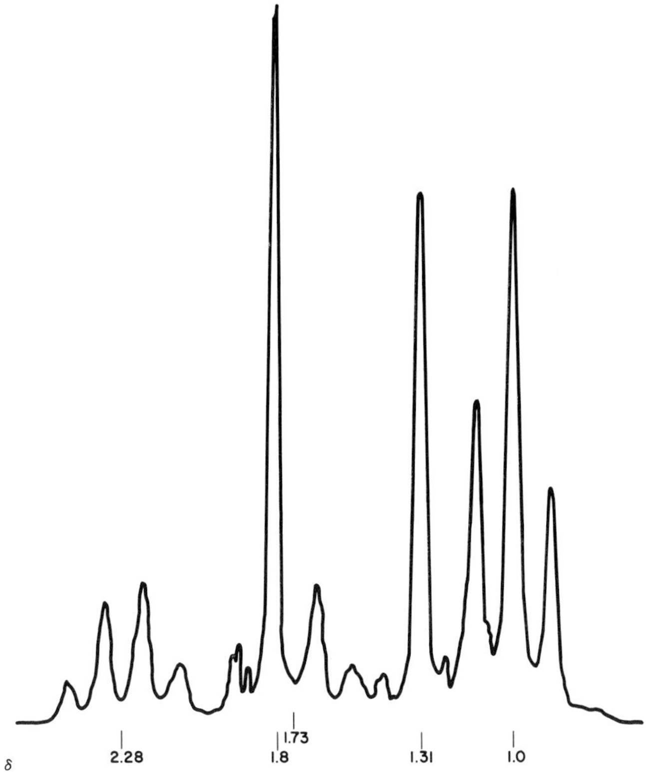

Figure 2. Low temperature (−40°C) NMR spectrum of methyl ethyl ketone diperoxide

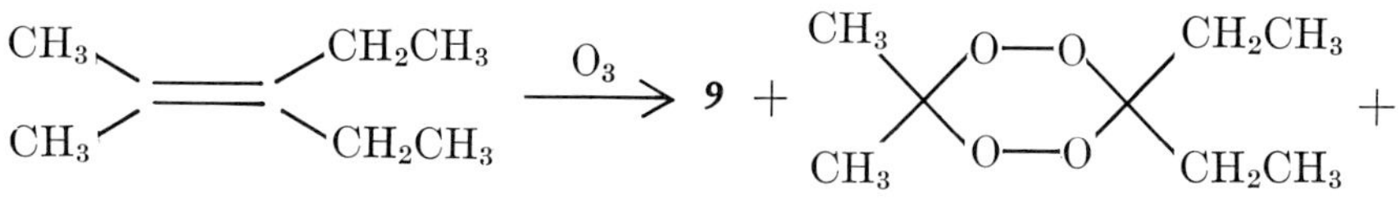

14 15

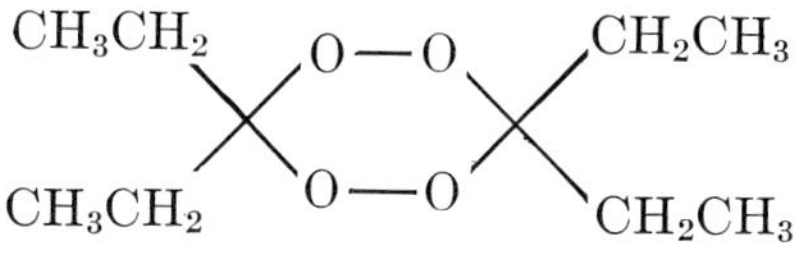

16

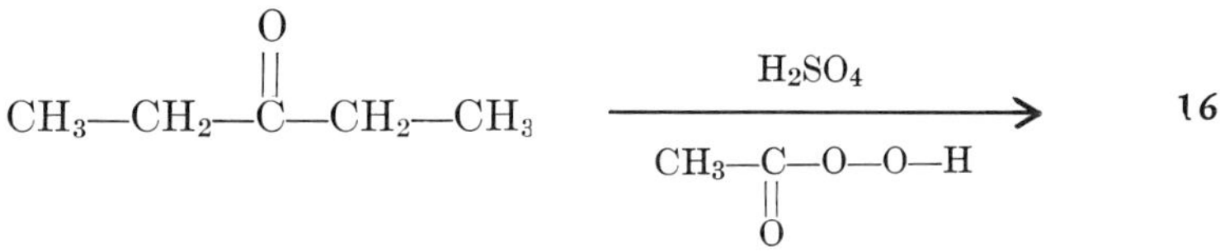

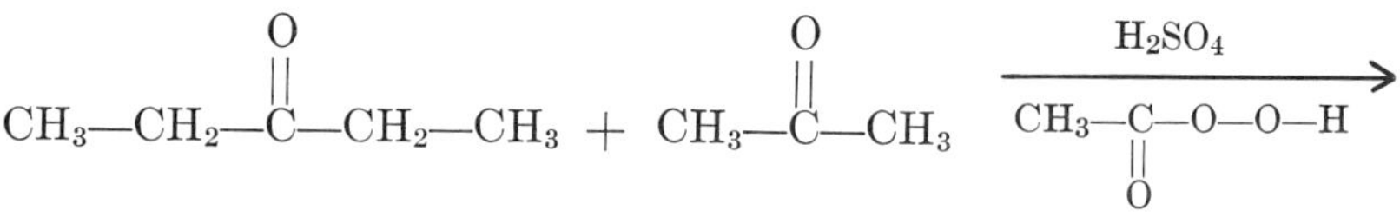

9 + 15 + 16

product also. Although the chemical shifts of the ring substituents in the cis and trans isomers can be different, the structures are so similar that any differences would probably not be detectable. The NMR results, therefore, are consistent with **11** as the cis or trans isomer or a mixture of both.

Since the same product or mixture of products was obtained from both cis and trans olefins (**8** and **13**), it is not possible to determine whether there is any influence of olefin stereochemistry on diperoxide stereochemistry in this case.

We hope to work with a stereoisomeric pair of olefins of type **2** where the substituent sizes are sufficiently different to permit an evaluation of the influence of olefin stereochemistry on the reaction. Similar considerations will have to be kept in mind when choosing a more complex olefin of type **3**.

We also have examined one final type of olefin. This type of olefin, represented by 2-methyl-3-ethyl-2-pentene, **14**, is also capable of giving cross diperoxides although in this case the diperoxide products are not capable of existing as stereoisomers. Ozonolysis of **14** gave three diperoxides, **9**, 1,1-dimethyl-4,4-diethyl-2,3,5,6-tetraoxacyclohexane, **15**, and 3-pentanone diperoxide, **16**.

Here again the peracetic oxidation method was used to prepare comparison samples of diperoxides. Oxidation of 3-pentanone, for example, gave a 46% yield of **16**. This method can also be used to prepare diperoxides such as **15**. When equimolar amounts of acetone and 3-pentanone are oxidized by the peracetic acid method, diperoxide **15** is obtained along with **9** and **16**. Similarly, oxidation of equimolar amounts of acetone and 2-butanone gave **9**, **10**, and **11**.

We have shown that cross diperoxides can be formed by various ozonolysis procedures. We now hope to parallel the work done where cross ozonides were produced—*i.e.,* to examine the influence of olefin stereochemistry and other reaction variables.

Literature Cited

1. Criegee, R., *Rec. Chem. Progr.* (1957) **18**, 111.
2. Riezebos, G., Grimmelikhuysen, J. C., Van Dorp, D. A., *Rec. Trav. Chim.* (1963) **82**, 1234.
3. Lorenz, O., Parks, C. R., *J. Org. Chem.* (1965) **30**, 1976.
4. Loan, L. D., Murray, R. W., Story, P. R., *J. Amer. Chem. Soc.* (1965) **87**, 737.
5. Privett, O. S., Nickell, E. C., *J. Amer. Oil Chemists' Soc.* (1964) **41**, 72.
6. Murray, R. W., Youssefyeh, R. D., Story, P. R., *J. Amer. Chem. Soc.* (1967) **89**, 2429.
7. Murray, R. W., Youssefyeh, R. D., Story, P. R., *J. Amer. Chem. Soc.* (1966) **88**, 3143.
8. Greenwood, F. L., *J. Amer. Chem. Soc.* (1966) **88**, 3146.
9. Bauld, N. L., Thompson, J. A., Hudson, C. E., Bailey, P. S., *J. Amer. Chem. Soc.* (1968) **90**, 1822.
10. Schröder, G., *Chem. Ber.* (1962) **95**, 733.
11. Kolsaker, P., *Acta. Chem. Scand.* (1965) **19**, 223.
12. Greenwood, F. L., Haske, B. J., *Tetrahedron Letters* (1965) 631.
13. Story, P. R., Murray, R. W., Youssefyeh, R. D., *J. Amer. Chem. Soc.* (1966) **88**, 3144.
14. Fliszar, S., Carles, J., *Canad. J. Chem.* (1969) **47**, 3921.
15. Criegee, R., Lohaus, G., *Ann.* (1953) **583**, 6.
16. Marvel, C. S., Nichols, V., *J. Org. Chem.* (1941) **6**, 296.
17. Pummerer, R., Schmidutz, G., Seifert, H., *Chem. Ber.* (1952) **85**, 535.
18. Criegee, R., Bath, S. S., Bornhaupt, B. V., *Chem. Ber.* (1960) **93**, 2891.
19. Criegee, R., Lohaus, G., *Chem. Ber.* (1953) **86**, 1.
20. Murray, R. W., Story, P. R., Loan, L. D., *J. Amer. Chem. Soc.* (1965) **87**, 3025.
21. Criegee, R., Korber, H., *Chem. Ber.* (1971) **104**, 1812.

22. Murray, R. W., Story, P. R., Kaplan, M. L., *J. Amer. Chem. Soc.* (1966) **88**, 526.
23. Lohringer, W., Sixt, J., U. S. Patent **3,076,852** (Feb. 5, 1963).
24. Jackman, L. M., "Applications of Nuclear Magnetic Resonance in Organic Chemistry," Section 7.2, Pergamon, New York, 1959.

RECEIVED May 20, 1971. Supported by the National Science Foundation grant No. GP-10895.

3

Fragmentation of Ozonides by Solvents

RUDOLF CRIEGEE and HELMUT KORBER

Institut für Organische Chemie, Universität Karlsruhe, Germany

Ozonides, bearing at least one H atom at the trioxolane ring, decompose quantitatively in some solvents into one molecule of acid and one molecule of aldehyde or ketone. Alcohols show the highest rate while solutions in hydrocarbons are stable. In methanol the reaction rates were measured at constant pH and salt concentration by automatic titration of the acid formed. The rates differ by a factor of 10^3, depending upon the constitution and the configuration of the ozonides. The abstraction of an H atom from the trioxolane ring occurs in the rate-determining step. A mechanism similar to that of Kornblum-DeLaMare is proposed. Several new ozonides were prepared. In three cases their configuration was established by resolution into optically active forms by chromatography on cellulose acetate.

Solutions of stilbene ozonides in methanol slowly decompose at room temperature with the formation of benzoic acid and benzaldehyde. Other ozonides show similar behavior. However, the reaction rates differ from solvent to solvent and depend on the constitution and configuration of the ozonide. Therefore, we decided to investigate these reactions more carefully. Many new ozonides—especially those with one or more aromatic substituents—had to be prepared, and their configurations had to be established.

Preparation of the Ozonides

Ozonides having one or two substituents at the trioxolane ring—in the latter case symmetrically or unsymmetrically—were prepared from the corresponding olefins by ozonization in pentane, usually at −78°C. The 3,5-disubstituted trioxolanes were separated into cis- and trans-isomers by crystallization or by column chromatography.

Table I. New Trioxolanes

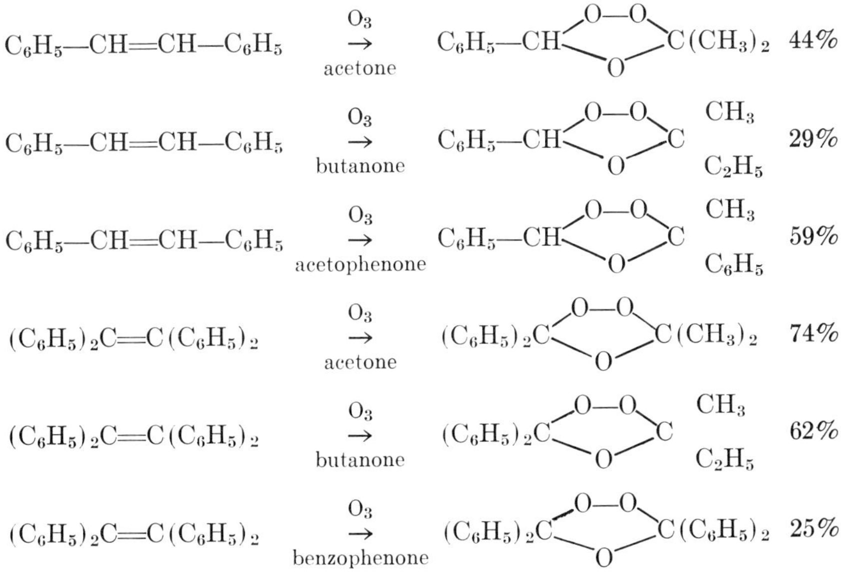

Some trisubstituted ozonides could be obtained only by the method of Murray, Story, and Loan (*1*)—namely, the ozonization of disubstituted olefins in ketones as solvents. Ozonization of tetraphenylethylene in acetone, methyl ethyl ketone, or benzophenone gave the hitherto unknown trioxolanes with four hydrocarbon residues (*2*). Table I shows some examples.

Other new ozonides include those of acenaphthylene and 3,4-dichloro-3,4-dimethyl-cyclobut-1-ene. They are the cyclic analogs of cis-stilbene ozonide and 1,4-dichloro-but-2-ene ozonide, respectively. Altogether, more than 45 ozonides were investigated.

Experimental

Ozone was prepared with an ozone generator (type OZ II of the Fischer Labortechnik, Bad Godesberg, Germany). Dry oxygen containing about 10% ozone was introduced at a speed of 10–20 liters/hour in the olefin solution at −78°C. Dry and olefin-free *n*-pentane was used as solvent.

After ozonation, the solvent usually was removed at 30°C under rotation. The residue was either distilled *in vacuo* or crystallized or purified by column chromatography on silica gel. *n*-Pentane, with increasing amounts of ether (up to 1.5%), served for the elution. Those mixtures of stereoisomeric ozonides which could not be separated preparatively were analyzed by their NMR spectra using a Varian A 60 spectrometer. The results are summarized in Tables II–V. More details can be found in Ref. 3.

Table II. Preparation and Properties

Starting Olefin	*1,2,4-Trioxolane*	*Formula*	*Yield, %*
Styrene	3-phenyl	$C_8H_8O_3$	71 [b]
p-Methoxystyrene	3-*p*-methoxyphenyl	$C_9H_{10}O_4$	41
p-Methylstyrene	3-*p*-tolyl	$C_9H_{10}O_3$	70
p-Chlorostyrene	3-*p*-chlorophenyl	$C_8H_7ClO_3$	70 [c]
p-Nitrostyrene	3-*p*-nitrophenyl	$C_8H_7NO_5$	44
Allyl chloride	3-chloromethyl	$C_3H_5ClO_3$	80
5-Chloropentene	3-(3-chloropropyl)	$C_5H_9ClO_3$	50

[a] Calculated values in parentheses.
[b] Plus 9% stilbene ozonide (cis/trans 4/6).

Table III. Preparation and Properties

Starting Olefin	*1,2,4-Trioxolane*	*Formula*	*Yield, %*
1,1-Diphenylethylene	3,3-diphenyl	$C_{14}H_{12}O_3$	57
2-Phenyl-but-1-ene	3-ethyl-3-phenyl	$C_{10}H_{12}O_3$	62
2-Phenyl-propene	3-methyl-3-phenyl	$C_9H_{10}O_3$	74 [b]
2-Methyl-propene	3,3-dimethyl	$C_4H_8O_3$	45 [c]
2,3-Dimethyl-but-1-ene	3-methyl-3-isopropyl	$C_6H_{12}O_3$	45

[a] Calculated values in parentheses.
[b] Plus 7% dimeric acetophenone peroxide.

of Monosubstituted Trioxolanes

Mp (Bp), °C/torr	*Analysis*[a]		*NMR Spectrum*[d] *in* CCl_4
	C	*H*	
(37-38/.01)			4.90(s), 4.85(s), 4.10(s), 2.70(m) 1:1:1:5
61-62	59.57 (59.37)	5.77 (5.53)	6.24(s), 4.73(s), 4.63(s), 4.06(s), 2.83 (AB system, $J = 9$Hz); 3:1:1:1:4
(55/0.01)	66.21 (66.05)	6.12 (6.07)	7.75(s), 4.81(s), 4.73(s), 4.08(s), 2.79 (AB system, $J = 9$Hz); 3:1:1:1:4
(62/0.03)	51.23 (51.49)	3.73 (3.38)	4.73(s), 4.67(s), 4.04(s), 2.60(s) 1:1:1:4
77	48.86 (48.74)	3.62 (3.58)	4.65(s), 4.61(s), 3.82(s), 2.02(AB system, $J = 9$Hz); 1:1:1:4
(51/18)	28.99 (28.94)	4.05 (4.05)	6.50(d, $J = 5$Hz), 5.05(s), 4.77(s), 4.60(t, $J = 5$Hz); 2:1:1:1
(36/0.04)	39.20 (39.36)	5.99 (5.95)	8.11(m), 6.45(m), 5.02(s), 4.90(s) 4:2:1:2

[c] Plus 13% *p,p'*-dichlorostilbene ozonide (cis/trans 4/6).
[d] Values of τ; TMS as internal standard.

of 3,3-Disubstituted Trioxolanes

Bp, °C/torr	*Analysis*[a]		*NMR Spectrum*[d] *in* CCl_4
	C	*H*	
n_D^{20} 1.5760	74.19 (73.67)	5.44 (5.30)	4.85(s), 2.75(m); 2:10
46/0.05	66.15 (66.65)	6.79 (6.71)	9.12(t, $J = 7$Hz), 8.02 (q, $J = 7$Hz), 4.99(s), 4.93(s), 2.67(m); 3:2:1:1:5
35/0.03	64.88 (65.05)	6.20 (6.07)	8.33(s), 4.94(s), 2.57(m); 3:2:5
44/143	45.62 (46.15)	7.61 (7.75)	8.60(s), 5.00(s); 6:2
48/37	54.31 (54.53)	9.23 (9.15)	9.06(d, J — 7Hz), 8.70(s), 8.06 (septet) 5.06(s), 4.95(s); 6:3:1:1:1

[c] Plus 4% dimeric acetone peroxide.
[d] Values of τ; TMS as internal standard.

Table IV. Preparation and Properties

Starting Olefin	*1,2,4-Trioxolane*	*Formula*	*Yield,[a] %*
trans-1-Phenylpropene	3-methyl-5-phenyl	$C_9H_{10}O_3$	54[c] (37/63)
trans-Crotyl bromide	3-methyl-5-bromo-methyl	$C_4H_7BrO_3$	74 (24:76)
1,4-Dibromo-but-2-ene (cis or trans)	3,5-bis(bromo-methyl)	$C_4H_6Br_2O_3$	55 (36:64)
trans-3,4-Dichloro-3,4-dimethyl-cyclo-butene		$C_6H_8Cl_2O_3$	61
trans-3,4-Dichloro-1,2,3,4-tetramethyl-cyclobutene		$C_8H_{12}Cl_2O_3$	43
Acenaphthylene		$C_{12}H_8O_3$	25

[a] Cis-trans ratio in parentheses.
[b] Calculated values in parentheses.

Results and Discussion

Configuration of cis-trans Isomeric Ozonides. In symmetrically disubstituted trioxolanes, the cis isomer is a meso and the trans isomer is a racemic form. Therefore, resolution into antipodes makes it possible to determine the configuration. In 1966, Loan, Murray, and Story (*4*) were able to obtain an optically active form of one of the diisopropyltrioxolanes, proving it to be the trans isomer. Their method—partial decomposition of the ozonide by brucine—could not be applied to the stilbene ozonides.

Chromatography of a benzene solution of the low melting stilbene ozonide (mp 92°C) on finely divided cellulose-2 1/2-acetate, however, yielded fractions of the ozonide with specific rotation at 360 nm of +8.94° and −20.45°, respectively (*5*). This isomer, therefore, is the trans compound. Under the same conditions, the higher melting ozonide (mp 100°C) gave no fraction with optical activity, in agreement with its cis configuration.

of 3,5-Disubstituted Trioxolanes

Mp, °C	*Analysis*[b] C	H	*NMR Spectrum*[d] *in* CCl_4
trans: 54 cis: liquid	65.08 (65.06)	6.24 (6.07)	cis: 8.57 (d, J = 5Hz), 4.52 (q, J = 5Hz), 3.97 (s), 2.60 (m); 3:1:1:5 trans: 8.60 (d, J = 5Hz), 4.49 (q, J = 5Hz), 4.00 (s), 2.60 (m); 3:1:1:5
trans: liquid cis: liquid	26.60 (26.25)	3.89 (3.86)	cis: 8.55 (d, J = 5Hz), 6.71 (d, J = 5Hz), 4.80 (q, J = 5Hz), 4.52 (q, J = 5Hz); 3:2:1:1 trans: 8.57 (d, J = 5Hz (, 6.64 (d, J = 4.5Hz), 4.60 (q, J = 5Hz), 4.65 (q, J = 4.5Hz); 3:2:1:1
trans: liquid cis: liquid	18.02 (18.34)	2.39 (2.31)	cis: 6.55 (d, J = 4.5Hz), 4.56 (t, J = 4.5Hz); 4:2 trans: 6.64 (d, J = 5Hz), 4.43 (t, J = 5Hz); 4:2
91	36.24 (36.21)	4.13 (4.05)	8.24 (s), 8.21 (s), 4.43 (s), 4.29 (s); 3:3:1:1
75	42.38 (42.33)	5.34 (5.29)	8.40 (s), 8.29 (s), 8.27 (s), 8.20 (s); 3:3:3:3
106	71.68 (72.00)	3.96 (4.03)	3.63 (s), 2.60 (m); 2:6

[c] Plus 15% stilbene ozonide; but-2-ene ozonide was also found by gas chromatography.
[d] Values of τ; TMS as internal standard.

In the same way, the 1,4-dichloro-but-2-ene ozonide with the shorter retention time on silica gel yielded fractions with $[\alpha]_{400} = -2.66°$ and $+4.2°$; the ozonide of 1,4-dibromo-2,3-dimethyl-but-2-ene, which only could be made in one form with a melting point of 42°C, yielded fractions with $[\alpha]_{330} = +50.7°$ and $-39.2°$. Both, therefore, have the trans configuration. It is remarkable that the trans ozonide in this case is the only product, even when pure *cis*-1,4-dibromo-2,3-dimethyl-but-2-ene is ozonized.

The configurations of the other ozonides are based on the assumption that the trans isomers have the lower dipole moments in comparison with the cis compounds. As a measure of the dipole moment the retention time in liquid phase or gas chromatography was used. The isomer with the lower retention time should be the trans compound.

Except for the trioxolanes with halo alkyl groups, the trans compounds always reacted more slowly with triphenylphosphine than did the cis isomers. The NMR values generally cannot be used to estimate the

Table V. Preparation and Properties

Starting Olefin	*Solvent, Temperature, °C*	*1,2,4-Trioxolane*	*Formula*
trans-Stilbene	acetone −78	3,3-dimethyl-5-phenyl	$C_{10}H_{12}O_3$
trans-Stilbene	acetophenone 0	3-methyl-3,5-diphenyl	$C_{15}H_{14}O_3$
cis-Stilbene	butanone −78	3-methyl-3-ethyl-5-phenyl	$C_{11}H_{14}O_3$
1,1-Diphenylpropene	pentane −78	3-methyl-5,5-diphenyl	$C_{15}H_{14}O_3$
2-Phenyl-but-2-ene	pentane −78	3,5-dimethyl-3-phenyl	$C_{10}H_{12}O_3$

[a] Calculated values in parentheses.
[b] Cis-trans ratio in parentheses.
[c] Plus 19% stilbene ozonide + 4% dimeric benzaldehyde peroxide.
[d] Phenyl and methyl in the cis position.

configuration because they are influenced by substituents in different ways. They can be of value, however, in series of homologous ozonides. As shown later, the rate of fragmentation is always higher for the trans ozonides, at least in the series of disubstituted trioxolanes.

Fragmentation Products. All ozonides except for the tetrasubstituted ones are cleaved quantitatively by many solvents. While symmetrically disubstituted ozonides give one mole of acid and one mole of aldehyde as products, and trisubstituted ozonides yield one mole acid and one mole of a ketone, in the case of monosubstituted or unsymmetrically 3,5-disubstituted trioxolanes, two kinds of acids and two kinds of aldehydes are possible. The fragmentation products in these cases were treated with diazomethane, and the mixture of the methyl esters was analyzed by gas chromatography. The results are shown in Table VI. In all of these examples, the reactions go predominantly in one direction. When R and R′ are more similar, however, both kinds of reaction products would be expected in more nearly equal amounts.

of Trisubstituted Trioxolanes

Yield,[b] %	Bp, °C/torr	Analysis[a] C	Analysis[a] H	NMR Spectrum[d] in CCl_4
44	39/0.04	67.35 (66.65)	6.62 (6.71)	8.49(s), 8.45(s), 3.96(s), 2.60(m); 3:3:1:5
29[c] (43:57)	trans: m.p. 54 cis: liquid	74.28 (74.36)	5.80 (5.82)	cis: 8.18(s), 4.05(s), 2.65 (m); 3:1:10 trans: 8.28(s), 3.98(s), 2.75(m); 3:1:10
59 (27:73)	70/0.02	68.34 (68.02)	7.31 (5.82)	cis:[d] 9.00(t, $J = 7$Hz), 8.50(s), 8.28(q, $J = 7$Hz), 4.08(s), 2.70(m); 3:3:2:1:5 trans: 9.04(t, $J = 7$Hz), 8.55(s), 8.16(q, $J = 7$Hz), 4.04(s), 2.70(m); 3:3:2:1:5
30[e]	94/0.02	74.18 (74.36)	5.82 (5.82)	8.74(d, $J = 5$Hz), 4.52(q, $J = 5$Hz), 2.76(m); 3:1:10
27[f] (42:58)	43/0.04	66.41 (66.65)	6.62 (6.71)	cis: 8.74(d, $J = 5$Hz), 8.37(s), 4.63(q, $J = 5$Hz), 2.67(m); 3:3:1:5 trans: 8.62(d, $J = 5$Hz), 8.31(s), 4.70(q, $J = 5$Hz), 2.67(m); 3:3:1:5

[e] Plus 32% benzophenone + 5% dimeric benzophenone peroxide.
[f] Plus 5% dimeric acetophenone peroxide.
[g] Values of τ; TMS as internal standard.

Table VI. Direction of Fragmentation of Unsymmetrical Ozonides

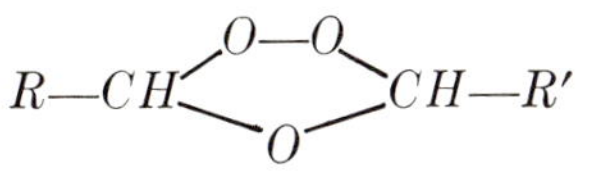

R	R′	Products, % RCO_2H	Products, % $R'CO_2H$
n-C_3H_7–	H	97	3
$ClCH_2CH_2CH_2$–	H	98	2
$ClCH_2$–	H	100	0
C_6H_5–	H	99	1
C_6H_5–	CH_3–	95	5

Rates of Fragmentation. The solvent dependence was studied first (Table IV). Solutions of styrene ozonide were stored at 50°C. At various intervals, samples were quenched with ice water, and the acid formed was titrated with 0.05*N* KOH, using phenolphthalein indicator. While the ozonide was stable in benzene, carbon tetrachloride, and chloroform solution, the rates of cleavage increased in oxygen- and nitrogen-containing solvents, methanol being the fastest solvent. As shown in Table VII, there is no clear connection between the rates of reaction and the polarities of the solvents, as expressed by the E_T values of Dimroth (*6*).

Table VII. Fragmentation of Styrene Ozonide at 50°C in Different Solvents

Solvent	$k_{50} \times 10^4$ *(sec⁻¹)*	E_T *(Ref. 6)*
Benzene	0	34.5
Diisopropylether	0.017	34.0
Acetonitril	0.038	46.0
Butanone	0.070	41.3
Dioxane	0.097	36
Tetrahydrofuran	0.47	37.4
Benzyl alcohol	1.83	50.8
Methanol	5.33	55.5

To study the dependence of the reaction rates on the constitution and configuration of the ozonides, an automatic method was worked out. Ozonides, dissolved in absolute methanol which was saturated with the barium salt of the acid produced in the reaction (mainly barium acetate or barium benzoate), were heated in a thermostat to 50°C. In a Combititrator (Methrom Co.) a methanol solution of barium methylate (0.1*N*) was added automatically at such speed that the pH of the solution remained unchanged. The variation in the volume of added methylate solution with reaction time was registered on chart paper. The reactions were carried out under pure nitrogen, and the curves always showed first-order kinetics.

From the kinetic results some rules can be established.

Rule 1: abstraction of a proton from the trioxolane ring occurs in the rate-determining step. This follows from the comparison of the rates

Table VIII. Isotope Effect

	$k_{50} \times 10^4$ *(sec⁻¹)*		
	H Ozonide	*D Ozonide*	k_H/k_D
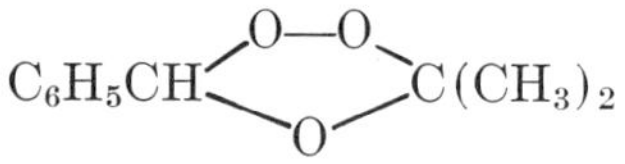	0.21	0.073	2.9
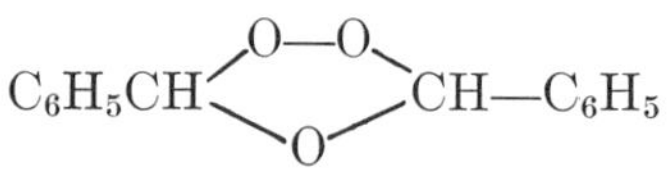	4.5	1.17	3.9
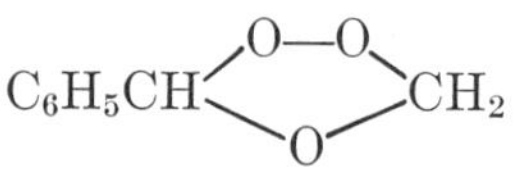	7.15	1.48	4.8

of protonated and deuterated ozonides (Table VIII). The isotope effect was found to be between 3 and 5. The deuterated ozonides had the deuterium labels in all positions of the trioxolane rings. They were prepared by ozonization of stilbene-d_2 in acetone or in pentane and of styrene-d_3 in pentane.

Rule 2: aromatic residues at the trioxolane ring render the removal of a hydrogen atom from the same carbon atom easier than does an aliphatic substituent. The fragmentation rates are 7 to 30 times higher in the first case (Table IX).

Table IX. Comparison of Aliphatic and Aromatic Substituents

	$k_{50} \times 10^4$ (sec^{-1})			
R =	$—C_3H_7$	$—CH_3$	$—C_6H_5$	*aromatic/ aliphatic*
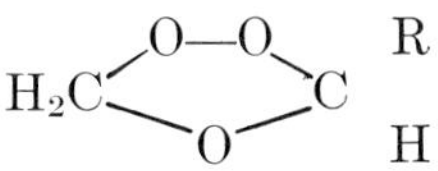	0.107		0.71	6.6
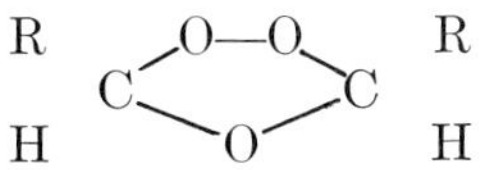	0.019		0.31	16
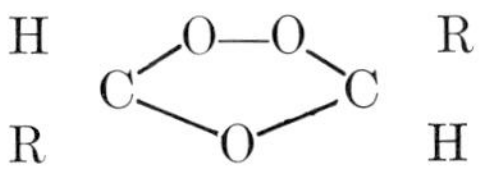	0.021		0.59	28
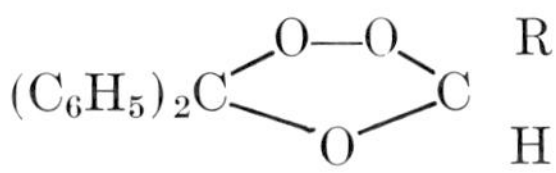		0.0083	0.083	10

Rule 3: electron-attracting substituents increase, electron-donating groups decrease the reaction rates. This is shown in Table X for para-substituted styrene ozonides. The Hammett equation gives a ρ-value of about $+0.9$.

Table X. Fragmentation of Substituted Styrene Ozonides in Methanol

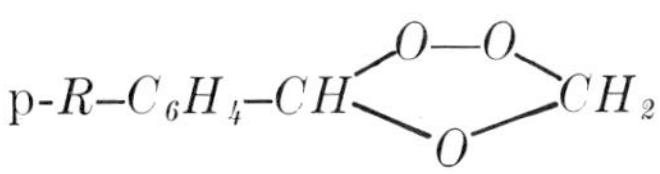		$k_{50} \times 10^4$ (sec^{-1})
R =	$CH_3O–$	4.75
	$CH_3–$	5.77
	H–	7.15
	Cl–	11.1
	$NO_2–$	53.7

Rule 4: in aliphatic substituents, halogen atoms show an inductive effect, decreasing with increasing distance of the halogen from the trioxolane ring (Table XI).

Table XI. Inductive Effect

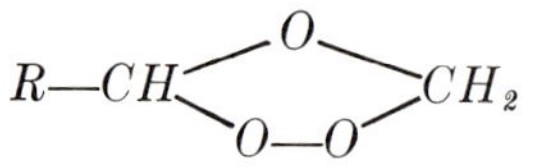

R =	$k_{50} \times 10^4$ (sec^{-1})
$CH_3CH_2CH_2$–	1.07
$ClCH_2CH_2CH_2$–	1.50
$ClCH_2$–	5.7

Rule 5: the release of strain by the fragmentation of the ozonide is very important to the rate of fragmentation. Table XII gives three pairs of ozonides. Each pair is chemically similar but differs in the Baeyer strain. The strained ozonides are cleaved up to more than 100 times faster than the unstrained ozonides.

Table XII. Influence of Strain

	$k_{50} \times 10^4$		$k_{50} \times 10^4$
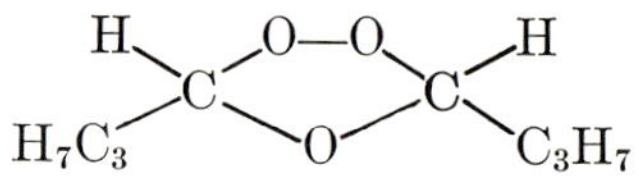	0.19	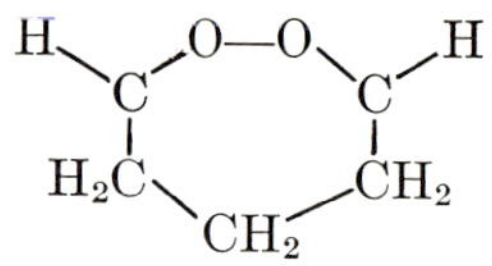	0.96
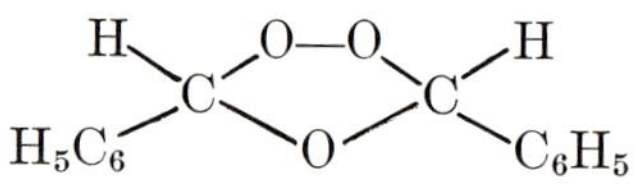	3.1	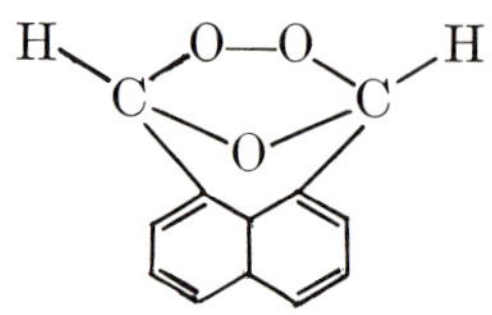	66.0
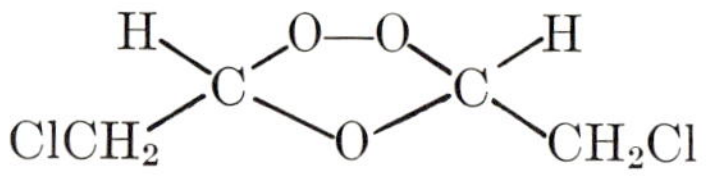	1.70	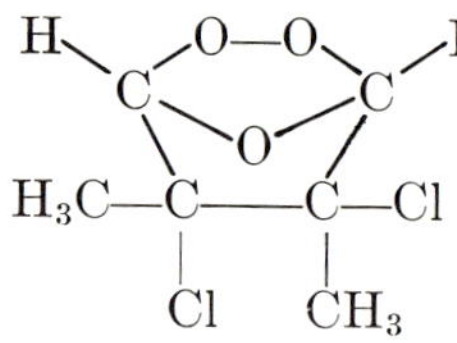	250

Rule 6: among stereoisomeric ozonides, the trans isomers (or the compounds which should be trans from the lower retention time) are always fragmentated faster than the cis isomers (Table XIII). The trans/cis factor, however, usually is not large. Only with halogen containing substituents does it exceed 10.

Table XIII. Reaction Rates of cis-trans Isomeric Ozonides

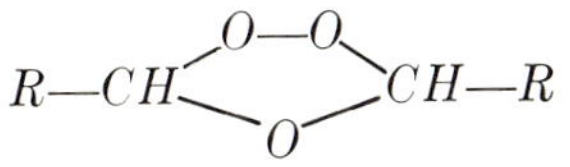

		$k_{50} \times 10^4$ (sec^{-1})		
		cis	*trans*	k_{trans}/k_{cis}
R =	CH_3	0.123	0.131	1.07
	n-C_3H_7–	0.193	0.213	1.15
	tert-C_4H_9	0.202	0.333	1.65
	C_6H_5	3.13	5.90	1.88
	$ClCH_2$–	1.70	19.5	11.5

Unfortunately, the dependence of the rates on pH cannot be measured by our method since acidimetric titrations are not possible in buffered solutions. However, in weakly acidic and neutral solutions, there cannot be much influence because otherwise no rate constants could have been obtained. In alkaline solution, on the other hand, a base induced fragmentation occurs, which is faster by some powers of 10 than the solvent fragmentation. Thus, in 0.025*N* methanolic NaOH at 0°C, the half lifetime of stilbene ozonide is only 2 minutes.

Mechanism of the Solvent Fragmentation of Ozonides. The driving force for the fragmentation, in any case, is the cleavage of the energy rich O—O bond of the ozonide and the release of conformational strain present in the trioxolane ring. However, the strongly exothermic fragmentation into stable products can occur only if there is an acceptor for the trioxolane proton. It is surprising that this acceptor need not be a strong base but can be a solvent with a heteroatom's possessing lone pairs of electrons.

The exceptional role of alcohols as solvents may arise from the fact that they are not only "bases" but simultaneously are "acids." Therefore, unlike the reaction with strong bases, free acids are formed instead of carboxylate anions.

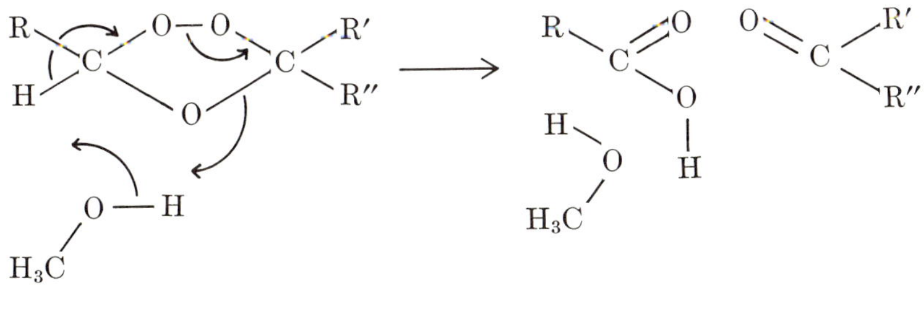

From our experiments we cannot say whether one or more molecules of methanol participates in the rate-determining step. The first possibility implies that one methanol molecule is simultaneously acceptor and donator of a proton, while in the second case these functions are carried out by different molecules. The fragmentation seems to be concerted because according to rules 1 and 5 both the abstraction of the proton and the release of strain are rate determining. Thus, we formulate a mechanism which is similar to the well known Kornblum-DeLaMare mechanism for the base induced decomposition of dialkyl peroxides (*7*).

While this mechanism is not in disagreement with the various kinetic results, it can explain only with difficulty all details of the reaction rates. The higher rates in the case of ozonides with electron-attracting substituents show that in the transition state there must be a partial negative charge at that C atom which loses the proton. Since, however, the rates are also influenced by steric factors (steric hindrance as well as steric effects upon the ring conformation), the whole situation is complicated. No effort has been made, therefore, to explain all rate differences, even small ones. From a practical standpoint the solvent fragmentation seems to offer a very mild method to cleave ozonides into acids and carbonyl compounds. In a recent paper (*8*) which describes the fragmentation of stilbene ozonides with dimethyl sulfoxide and dimethyl formamide the authors propose a similar mechanism.

Literature Cited

1. Murray, R. W., Story, R. P., Loan, L. D., *J. Amer. Chem. Soc.* (1965) **87,** 3025.
2. Criegee, R., Korber, H., *Chem. Ber.* (1971) **104,** 1812.
3. Korber, H., Ph.D. thesis, Universitat Karlsruhe, 1970.
4. Loan, L. D., Murray, R. W., Story, P. R., *J. Amer. Chem. Soc.* (1965) **87,** 737.
5. Criegee, R., Korber, H., *Chem. Ber.* (1971) **104,** 1807.
6. Dimroth, K., Reichardt, C., Siepmann, T., Bohlmann, F., *Ann. Chem.* (1963) **661,** 1.
7. Kornblum, N., DeLaMare, H. E., *J. Amer. Chem. Soc.* (1951) **73,** 880.
8. Ellam, R. M., Padbury, J. M., *Chem. Commun.* (1971) 1094.

RECEIVED May 20, 1971.

4

Quantitative Investigation of the Ozonolysis Reaction

XVI. Substituent Effects in the Ozonization Rates of Ethylenes

JACQUES CARLES and SÁNDOR FLISZÁR

University of Montreal, Montreal, Québec, Canada

The effects contributed by alkyl groups to the relative rate constants, k_{rel}*, for the reaction of ozone with cis- and trans-1,2-disubstituted ethylenes are adequately described by Taft's equation* $k_{rel} = k^\circ_{rel} + \rho\Sigma\sigma^*$*, where* $\Sigma\sigma^*$ *is the sum of Taft's polar substituent constants. The positive* ρ^* *values (3.75 for trans- and 2.60 for cis-1,2-disubstituted ethylenes) indicate that for these olefins the rate-determining step is a nucleophilic process. The results are interpreted by assuming that the electrophilic attack of ozone on the carbon–carbon double bond can result either in a 1,3-dipolar cycloaddition (in which case the over-all process appears to be electrophilic) or in the reversible formation of a complex (for which the ring closure to give the 1,2,3-trioxolane is the nucleophilic rate-determining step).*

Most of the experimental studies concerning the mode of attack of ozone on carbon–carbon double bonds have been done with aromatic compounds. From the ozonolysis of polycyclic aromatic compounds, Badger (*1*) and Brown (*2, 3*) concluded that the ozone makes a one-step attack on the unsaturated system at the bond with the lowest bond localization energy, rather than an attack involving first the atom with the lowest atom localization energy. This is supported by the high yield (71%) of 4-formyl-5-phenanthrene carboxylic acid which is produced in the ozone attack on the 1:2 bond of pyrene (*1*)—the bond with the lowest bond localization energy.

On the other hand, studies on the ozonolysis of substituted benzenes by van Dijk (*4*) and Wibaut (*5, 6*) seem to indicate that the ozone attack is electrophilic in nature; the results presume an attack at the position with the lowest atom localization energy. This is confirmed by Bailey's finding (*7*) that anthracene is attacked by ozone at the positions with the lowest atom localization energy—*i.e.*, 9 and 10 to give anthraquinone in 70% yield.

More recent studies by Nakagawa, *et al.* (*8*) confirm the interpretation of the ozone attack on carbon–carbon double bonds in terms of an electrophilic attack. Their investigations on the kinetics of ozonization of polyalkylbenzenes, in CCl_4 and CH_3COOH solution, indicate that the logarithms of the rate constants for the ozonolysis of polymethylbenzenes increase linearly with the number of methyl substituents on the aromatic nucleus.

Similar conclusions are drawn by Cvetanović *et al.* from their results of ozonization of alkenes in the gas phase (*9*) and in CCl_4 solution (*10*). The rate constants for the ozonolysis of chloroethylenes and allyl chloride, in CCl_4 solution, indicate (*11*) that the rate of ozone attack decreases rapidly as the number of chlorine atoms in the olefin molecules is increased. However, to explain the departures from simple correlations, in some cases steric effects and the dipolar character of ozone had to be invoked (*10*). The relevance of the dipolar character of ozone in its reactions has also been stressed by Huisgen (*12*), who provided evidence that the ozone–olefin reaction is usually a 1,3-dipolar cycloaddition.

Finally (*13*), the relative rate constants for the reaction of ozone with selected ring-substituted styrenes in CCl_4 solution indicate that the second-order rate constants obey Hammett's equation: $\log k = \log k_0 + \rho\sigma$. The negative value of ρ (-0.91 ± 0.03) confirms that for these olefins the ozone attack is electrophilic in nature.

The trends in the ozonolysis rates of simple olefins require further examination. The object of the present work was to obtain kinetic data on the reaction of ozone with simple alkyl-substituted ethylenes. The results will allow a discussion of the substituent effects on the reaction rates and of the mode of ozone attack on the carbon–carbon double bond.

Relative Rate Constants

On the basis of the evidence (*10*) that, in solution, the ozone–olefin reaction takes place with a 1:1 stoichiometry, the consumption of olefin i may be written

$$-dx_i/dt = k_i[O_3]x_i \tag{1}$$

where x_i is the concentration of olefin i at time $t = t$, and k_i the rate constant for its reaction with ozone. As indicated by the results obtained in a previous study (*10*), the integral rate law,

$$\log(x[O_3]_o/[O_3]x_o) = kt(x_o - [O_3]_o)/2.3 \tag{2}$$

which is derived from Equation 1, satisfactorily describes the kinetics of a variety of olefins with ozone in CCl_4 solution. In Equation 2, $[O_3]_o$ and x_o are the ozone and olefin initial concentrations and it is assumed that $[O_3]_o \neq x_o$. The agreement of Equation 2 with the experimental results thus confirms the order of reaction postulated in Equation 1. Attempts to calculate the rate constants, made with the assumption of orders of reaction different from unity (referred to each reactant), failed in giving constant values for k.

In the present work, ozone was continuously bubbled through the olefin solution at a constant flow rate. Assuming a steady ozone concentration $[O_3]$ under these conditions, the integral rate law which is derived from Equation 1 is as follows:

$$\ln(x_i/x_{i,o}) = -k_i[O_3]t \tag{3}$$

where $x_{i,o}$ is the concentration of olefin i at time $t = 0$.

Hence, in the conventional ozonolysis of a mixture of two olefins which are identified by the subscripts 1 and 2, it is

$$\begin{aligned} \log(x_1/x_{1,o}) &= -k_1[O_3]t/2.3 \\ \log(x_2/x_{2,o}) &= -k_2[O_3]t/2.3 \end{aligned} \tag{4}$$

and, consequently,

$$k_1/k_2 = \log(x_1/x_{1,o})/\log(x_2/x_{2,o}) \tag{5}$$

This equation enables relative rate constants

$$k_{rel} = k_1/k_2 \tag{6}$$

(with respect to olefin 2) to be measured in a simple way, by conventional ozonolysis, from the analyses of the unreacted olefins. The fact that k_{rel} is independent of the ozone concentration and of the reaction time t (as predicted by Equation 5) considerably facilitates the measurements of k_{rel}.

The results in Table I concern the progressive ozonolysis of a CCl_4 solution of p-methylstyrene (1) and styrene (2), at 0°C, where the initial concentrations are $x_{1,o} = 0.05M$ and $x_{2,o} = 0.1M$. The constancy of k_{rel} which is observed clearly confirms that k_{rel} is independent of reaction time.

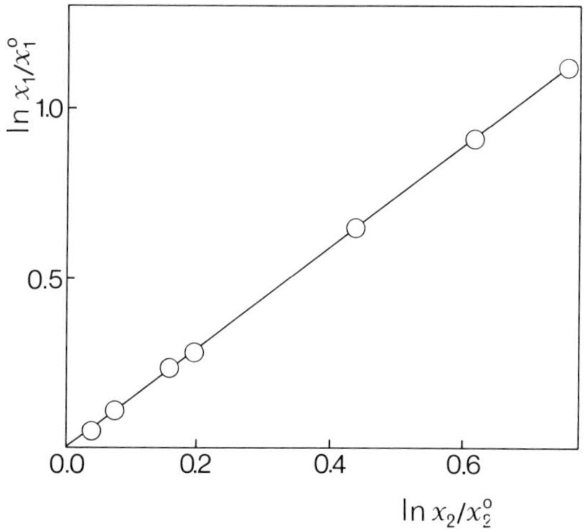

Figure 1. Typical plot for calculating k_{rel}: p-*methylstyrene (1)* vs. *styrene (2)*

To improve the analysis of the experimental results, it appears adequate to plot $\log(x_1/x_{1,o})$ *vs.* $\log(x_2/x_{2,o})$, as indicated in Figure 1, and to calculate the slope—*i.e.*, k_{rel}, by a least-squares method.

Table II presents a summary of the values of k_{rel} which were obtained in the way just indicated, for *p*-methylstyrene, in CCl_4 solution, with respect to styrene. These results indicate that k_{rel} is invariant with respect to the ozone flow rate, the temperature of ozonolysis, and the olefin initial concentration. The fact that k_{rel} is independent of temperature in this case is readily understood when it is considered that the experimental activation energies for the ozone–olefin reaction are likely to be very similar for *p*-methylstyrene and styrene.

Table I. Relative Rate Constants for the Ozonolysis of *p*-Methylstyrene (1) with Respect to Styrene (2)

	Olefin Concentration, mole/liter		
Ozonolysis, % [a]	x_1	x_2	$k_{rel} = k_1/k_2$
0	0.0500	0.1000	—
5	0.0474	0.0966	1.54
10	0.0450	0.0933	1.52
20	0.0397	0.0856	1.48
25	0.0375	0.0826	1.50
48	0.0259	0.0648	1.52
67	0.0164	0.0470	1.48
			Av. 1.506 ±0.025

[a] With respect to $x^o_1 = 0.0500$.

Table II. Relative Reaction Rates for *p*-Methylstyrene (1) with Respect to Styrene (2), in CCl_4 Solution, under Various Conditions

O_3 *Output, mmole/min*	*Temperature, °C*	$x°_1$ *(M)*	$x°_2$ *(M)*	k^a_{rel}	*n*
0.090	0	0.10	0.10	1.48	4
0.120	0	0.10	0.10	1.51	3
0.142	0	0.10	0.10	1.52	3
0.142	35	0.10	0.10	1.52	2
0.142	0	0.05	0.05	1.49	3
0.142	0	0.025	0.025	1.49	2
0.114	0	0.05	0.10	1.51	6
				Av. 1.503	

[a] Average values, from *n* independent determinations.

Experimental

Materials. A sample of pure *cis*-di-*tert*-butylethylene was prepared by R. Criegee (Karlsruhe). All other olefins are commercial products (K & K Laboratories Inc. and Aldrich Chemical Co. Inc.); their purity was controlled by gas chromatographic analysis.

Ozonolysis Procedure. Ten ml of olefin solution were ozonized at 0°C with an oxygen–ozone mixture. The ozone concentration was sufficiently high ($\approx$ 4%) to make the ozonolyses possible with relatively low O_2–O_3 flow rates ($\approx$ 0.15 mmole O_3/min) in order to prevent olefin loss by evaporation. Control experiments, performed with pure O_2, indicated that no appreciable amount of olefin is lost by bubbling the gas in the olefin solution in time intervals comparable with those required for the actual experiments.

Gas Chromatographic Analyses. Were performed with a 5750 F & M gas chromatograph, using a flame detector. Six-foot columns of silicone rubber UCW98 were used for styrene, the ring-substituted styrenes, 3-phenylpropene, *cis*- and *trans*-stilbene, and *cis*-di-*tert*-butylethylene. Six-foot columns of 10% Carbowax 20M were used for styrene, 3-iodopropene, 3-chloropropene, *cis*- and *trans*-3-hexene, *cis*- and *trans*-4,4-dimethyl-2-pentene, *cis*- and *trans*-5,5-dimethyl-3-hexene, *cis*- and *trans*-4-methyl-2-pentene, *trans*-2-methyl-3-hexene, *cis*- and *trans*-2,5-dimethyl-3-hexene, *trans*-di-*tert*-butylethylene, and 3-ethoxypropene. A 20-foot column of 10% nitrilesilicone on Chromosorb P was used for *cis*- and *trans*-4-octene and 3-bromopropene. A 6-foot 20% propylene carbonate column was used for *cis*- and *trans*-2-hexene, *cis*- and *trans*-2-pentene, and 1-pentene. A 20% diisopropyl phthalate 6-ft column was used for 3,3-dimethyl-1-butene and 3-methyl-1-butene. All analyses were repeated at least three times. The areas under the peaks corresponding to the olefins were measured by a planimeter and compared with the areas which were obtained under the same conditions with adequate reference solutions.

Results and Discussion

In the following discussion, the relative rate constants (with respect to styrene) are calculated by a least-squares analysis of the data, using Equation 5, from six to 10 experimental results. The standard deviations (s) from the regression line (k_{rel}) are also indicated.

The results obtained for the cis- and trans-1,2-disubstituted ethylenes are indicated, for the symmetrical and the unsymmetrical olefins, respectively, in Tables III and IV and in Table V for the monosubstituted ethylenes.

A comparison of the results obtained with alkyl-substituted ethylenes (Tables III and IV) indicates that k_{rel} decreases with increasing bulk of the substituents. This is also found to be true with the terminal olefins (Table V), as indicated by the lowering of k_{rel} from 1.11 to 0.407 when *n*-propyl is replaced by the *tert*-butyl group. The terminal olefins bearing an electron-withdrawing substituent also indicate a significant decrease of k_{rel}, as compared with 1-pentene, which is in agreement with Cvetanović's observations (*11*).

From this superficial consideration of substituent effects it would thus appear that the ozone attack is electrophilic in nature and that important steric effects contribute in determining the rate constants. The influence of steric effects could be adequately illustrated by the

Table III. Relative Rate Constants for the Ozonolysis of Symmetrical Olefins (RCH=CHR) in CCl_4

	cis		*trans*	
R	k_{rel}	s	k_{rel}	s
C_2H_5	2.13	0.14	2.98	0.15
n-C_3H_7	1.150	0.044	2.05	0.32
i-C_3H_7	0.471	0.014	0.613	0.026
tert-C_4H_9	0.197	0.029	0.088	0.004
C_6H_5	0.233	0.016	3.13	0.495

Table IV. Relative Rate Constants for the Ozonolysis of Unsymmetrical Olefins ($R_1CH{=}CHR_2$) in CCl_4

		cis		*trans*	
R_1	R_2	k_{rel}	s	k_{rel}	s
CH_3	C_2H_5	3.44	0.09	4.99	0.06
CH_3	*n*-C_3H_7	2.33	0.09	4.25	0.16
CH_3	*i*-C_3H_7	1.87	0.05	2.59	0.08
CH_3	*tert*-C_4H_9	0.932	0.041	0.927	0.051
C_2H_5	*i*-C_3H_7	([a])	([a])	1.26	0.02
C_2H_5	*tert*-C_4H_9	0.48	0.02	0.45	0.02

[a] Not determined.

Table V. Relative Rate Constants for the Ozonolysis of Monosubstituted Ethylenes in CCl_4

Olefin	k_{rel}	s
n-$C_3H_7-CH=CH_2$	1.11	0.05
i-$C_3H_7-CH=CH_2$	0.680	0.014
tert-$C_4H_9-CH=CH_2$	0.407	0.016
$C_6H_5CH_2-CH=CH_2$	0.113	0.008
$C_2H_5OCH_2-CH=CH_2$	0.177	0.006
$ICH_2-CH=CH_2$	0.174	0.006
$BrCH_2-CH=CH_2$	0.148	0.004
$ClCH_2-CH=CH_2$	0.205	0.006

results in Table III, which indicate that k_{rel} drops from 2.98 to 0.008, in trans-1,2-disubstituted ethylenes when the ethyl groups are replaced by two *tert*-butyl groups.

It appears, however, that such a simplified explanation, in which the effects from the alkyl substituents are interpreted mainly on steric grounds, meets with difficulties. The relative rate constants of ozonation can, in fact, differ considerably for ethylenes which have different substituents although their steric factors are similar. This is clearly indicated by comparing the results obtained for 1-pentene and iodo-3-propene. The n-propyl and CH_2I groups exhibit similar steric effects, as indicated by Taft's steric constants (E_s = −0.36 for n-C_3H_7; E_s = −0.37 for CH_2I); yet k_{rel} is about six times greater for 1-propene than for iodo-3-propene. Thus, it appears that the polar effects contributed by the groups to k_{rel} need a closer examination.

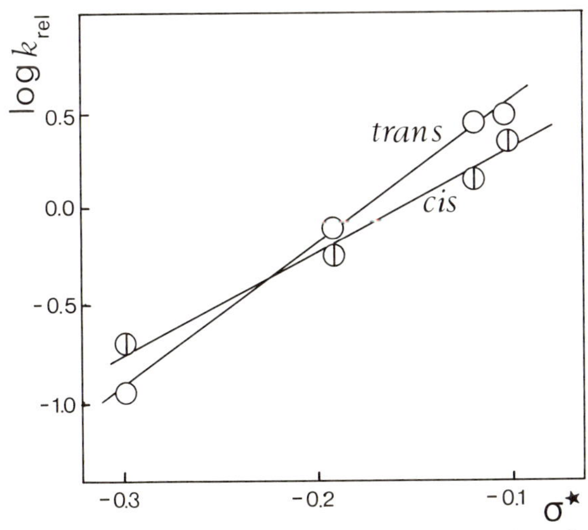

Figure 2. Comparison of log k_{rel} with Taft's polar σ constants for symmetrical 1,2-disubstituted ethylenes*

The participation of polar effects can be studied adequately by Taft's equation (*14*):

$$\log(k/k^{\circ}) = \rho^*\sigma^* \tag{7}$$

where k is the rate constant for an olefin $R_1CH{=}CHR$, and k° is the rate constant when $R = CH_3$. In terms of relative rate constants (Equation 6), this equation becomes

$$\log k_{rel} = \log k^{\circ}_{rel} + \rho^*\sigma^* \tag{8}$$

The satisfactory agreement of Equation 8 with the experimental results given in Tables III–V is shown in Figures 2-5. The good

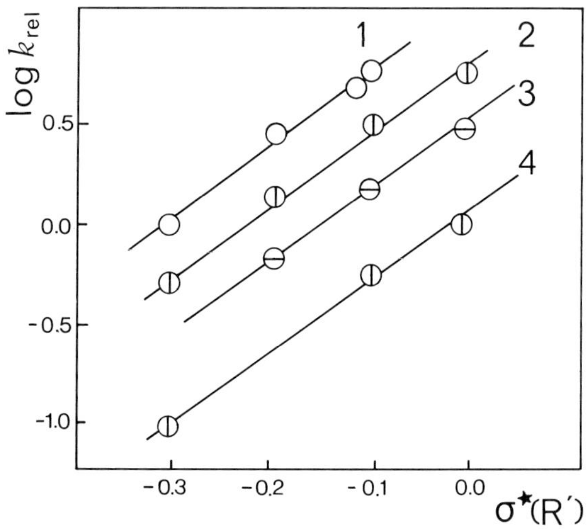

Figure 3. Verification of Equation 8 for trans-*1,2-disubstituted ethylenes:* $CH_3CH{=}CHR$ *(1),* $C_2H_5CH{=}CHR$ *(2),* i-$C_3H_7CH{=}CHR$ *(3), and* tert-$C_4H_9CH{=}CHR$ *(4)*

correlations which are observed seem to indicate an important contribution of polar effects and a minor (if any) influence of steric effects on k_{rel}. This point is supported by an evaluation of the relative importance of polar and steric effects by Taft's generalized equation (*15*):

$$\log(k/k^{\circ}) = \rho^*\sigma^* + \delta E_s \tag{9}$$

which, in terms of relative rate constants is:

$$\log k_{rel} = \log k^{\circ}_{rel} + \rho^*\sigma^* + \delta E_s \tag{10}$$

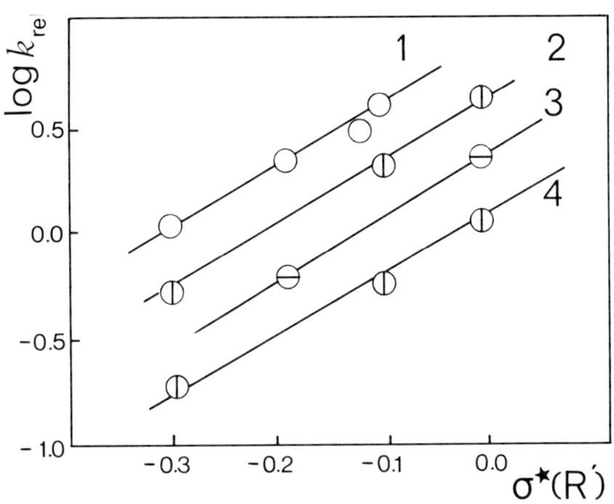

Figure 4. Verification of Equation 8 for cis-1,2-disubstituted ethylenes: $CH_3CH{=}CHR$ *(1),* $C_2H_5CH{=}CHR$ *(2),* i-$C_3H_7CH{=}CHR$ *(3), and* tert-$C_4H_9CH{=}CHR$ *(4)*

The symmetrical cis- and trans-olefins RCH═CHR (Table III) are particularly suited to this evaluation by Equation 10 because the polar reaction constant ρ^* and the steric correlation coefficient δ reflect, in this series of olefins, the effect of varying two substituents. Hence, the expected effects (both on ρ^* and δ) are more pronounced than those expected for olefins in which only one substituent is varied.

A least-squares calculation of ρ^* and δ by Equation 10, from the data in Table III, gives the following results for cis- and trans-olefins, RCH═CHR.

$$\text{trans} \quad \rho^* = 7.58 \quad \delta = 0.019$$

$$\text{cis} \quad \rho^* = 5.37 \quad \delta = 0.0027$$

When Equation 10 is applied to symmetrical olefins, the term k°_{rel} refers to 2-butene (R = CH_3). The calculated k°_{rel} values are: 17.5 for *trans*-2-butene and 6.13 for the cis-isomer.

The above calculations show that the steric contributions are small compared with the important polar effects, both in trans- and cis-olefins. Hence, in the following discussion, attention is focused on the predominant polar effects contributed by substituents, although a minor steric effect can possibly influence the overall reaction scheme.

Figure 3 indicates that for the trans-olefins $CH_3CH{=}CHR$, $C_2H_5CH{=}CHR$, *i*-$C_3H_7CH{=}CHR$, and *tert*-$C_4H_9CH{=}CHR$ the slopes (Equation 8) are similar, within experimental error. The same observation also applies to the cis-olefins represented in Figure 4. This suggests that in olefins $R_1CH{=}CHR_2$, the substituent effects possibly obey a simple additive rule—*i.e.*,

$$\log(k/k^\circ) = \rho^*\sigma^*(R_1) + \rho^*\sigma^*(R_2) \qquad (11)$$

Hence, in terms of relative rate constants, the dependence of k_{rel} on polar effects contributed by both substituents is adequately described by the following equation:

$$\log k_{rel} = \log k^\circ_{rel} + \rho^*\Sigma\sigma^* \qquad (12)$$

where k°_{rel} is the relative rate constant of *trans*- (or *cis*-) 2-butene. Verification of Equation 12 by the experimental results in Tables III and IV is shown in Figures 6 and 7.

For symmetrical 1,2-disubstituted ethylenes $\sigma^*(R_1) = \sigma^*(R_2)$ and hence, $\rho^*\Sigma\sigma^* = 2\rho^*\sigma^*$. The apparent slope (Figure 2) for symmetrical olefins is, thus, twice that observed for unsymmetrical olefins, or twice that observed for unsymmetrical olefins or twice that calculated by Equation 12.

Perhaps the most interesting point which emerges from the results is that in ethylenes bearing electron-releasing alkyl substituents the rate-determining step appears to be a nucleophilic process, as indicated by the positive ρ^* values. This does not contradict the assumption that the first step in the ozone–olefin reaction is an electrophilic attack of ozone on the carbon–carbon double bond. The present observations also agree with some of the results obtained recently by Pritzkow *et al.* (*16*) for alkyl mono-substituted ethylenes in ethanol solution at −60°C.

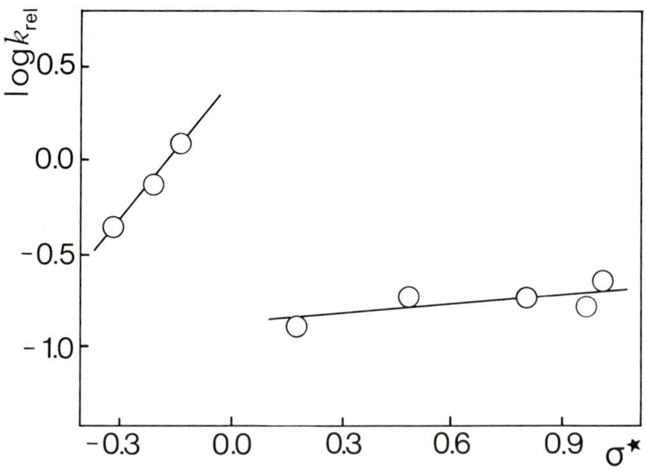

Figure 5. Verification of Equation 8 for $RCH{=}CH_2$ *olefins*

Figure 5 indicates a discontinuity in the Taft plots between the mono-substituted ethylenes bearing electron-releasing groups and those bearing electron-withdrawing groups. It is this discontinuity which led to the conclusion that the ozone attack is electrophilic in nature. In fact, it now appears that this discontinuity can be interpreted in terms of a change in mechanism.

The present results can be explained simply by considering that the ozone attack can proceed (a) *via* a 1,3-dipolar cycloaddition (*12*), and (b) *via* the formation of a π- or σ-complex (Scheme I). The possible occurrence of such complexes has already been suggested by Bailey (*17-19*), Murray (*20*), and Cvetanović (*11*).

Scheme I

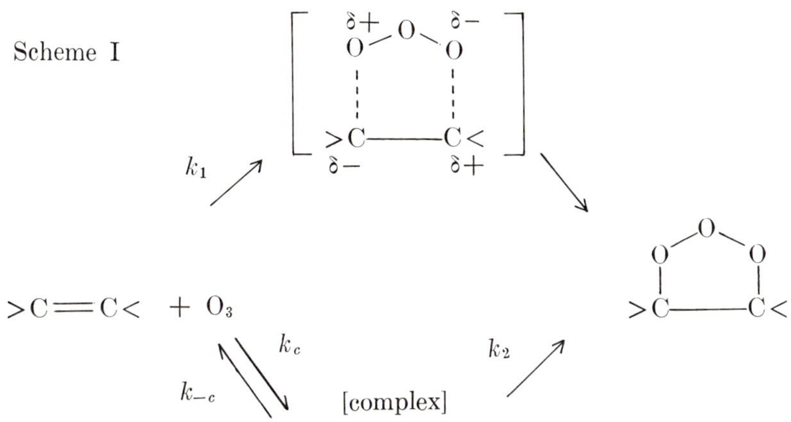

The following structures were suggested for the intermediary complex:

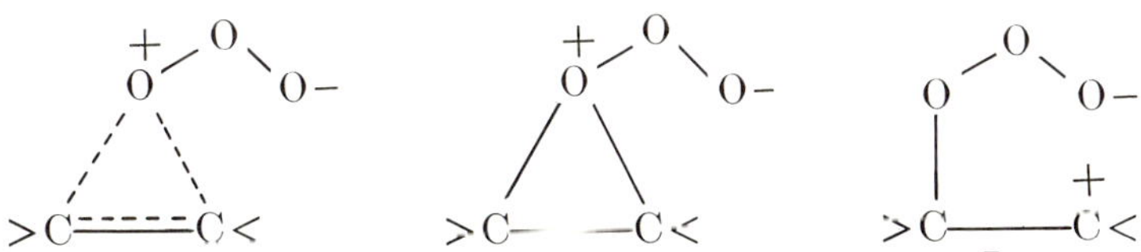

Ia Ib Ic

The contribution of Ic is not considered likely; the possible rotation about the carbon–carbon σ bond would result in a non-stereospecific ozone addition on the olefin—in violation of Greenwood's observations (*21, 22*) that a trans- (or cis-) olefin yields only trans- (or cis-) primary ozonide. Retention of configuration is also shown in those cases in which epoxide is formed (*20*). Finally, a recent communication by Bailey *et al.* (*23*) reports the formation of free radicals, arising from the homolytic dissociation of a π or σ complex, in the ozonolysis of terminal olefins with bulky substituents capable of stabilizing free radicals.

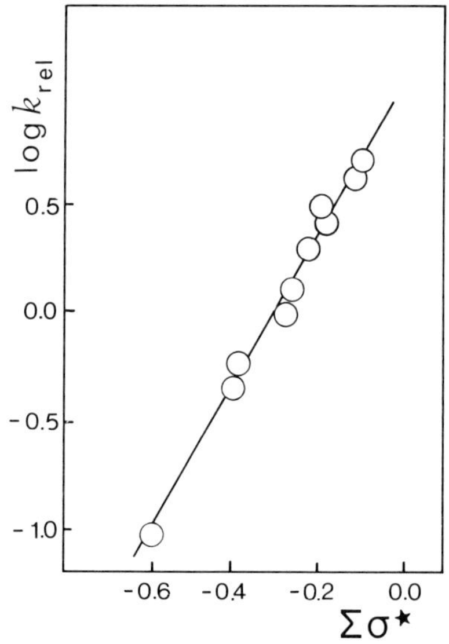

Figure 6. Verification of Equation 12 for trans-1,2-disubstituted ethylenes

In line with the suggestions made earlier by Cvetanović *et al.* (*11*) it is assumed that the complex is in equilibrium with the reactants. No accumulation of complex in the ozonolysis mixtures has ever been observed; hence, the assumption of a stationary state for the complex appears reasonable. Consequently, from a consideration of the processes indicated in Scheme I, the rate constant k for the reaction of ozone with olefin is:

$$k = k_1 + k_2K_c/(1+k_2/k_{-c}) \tag{13}$$

where $K_c = k_c/k_{-c}$. By assuming that $k_2 << k_{-c}$, as suggested by Cvetanović (*11*), the following approximate solution for k is obtained:

$$k = k_1 + k_2K_c \tag{14}$$

The process of ozone cycloaddition (path 1) implies postulates similar to those discussed by Huisgen (*12*) in terms of a 1,3-dipolar cycloaddition. Although the extent of simultaneity in the formation of the two C–O bonds is an open question, it is assumed that the transition state closely resembles the final state—the primary ozonide—and that its final conversion to give the primary ozonide occurs rapidly. The rate-determining step is thus the addition of ozone on the olefin. The electrophilic tendency of ozone, which is shown in several cases to play a dominant role in ozonolysis and the proposal of a 1,3-dipolar cycloaddition may well be accommodated within this scheme.

Pathway 2 (Scheme I) indicates that electron-releasing substituents —*e.g.*, the alkyl groups—are also likely to promote the formation of the π (or σ) complex—*i.e.*, to increase the value of K_c. In contrast to the situation in which the overall electrophilic character of the reaction is preserved because the ozone addition occurs *via* a 1,3-dipolar cyclic mechanism in the sense discussed above, the effect of an important electron release is considered to retard the ring closure—*i.e.*, the attack of O^- on the corresponding C atom. This would result in reducing the rate of conversion of the intermediate ozone–olefin adduct to give the primary ozonide. Hence, according to Scheme I, when the rate-determining step is no longer the formation of the intermediary ozone–olefin adduct but its conversion into primary ozonide, the overall process is nucleophilic in nature provided that the final step is the ring-closure involving O^-. [This interpretation offers a justification for Cvetanović's hypothesis (*11*) that $k_2 << k_{-c}$.]

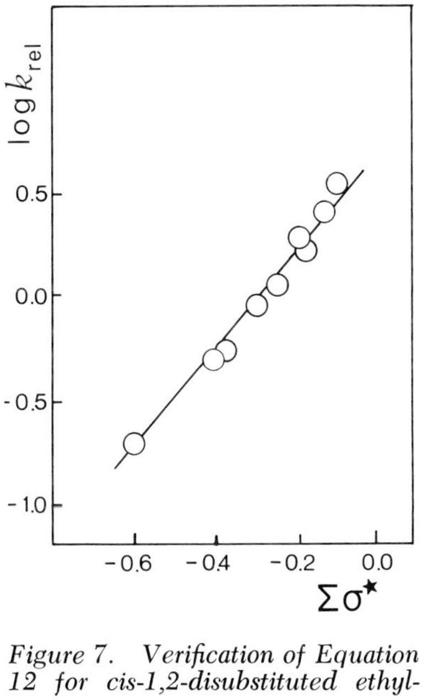

Figure 7. Verification of Equation 12 for cis-1,2-disubstituted ethylenes

Conversely, the results obtained for the monosubstituted ethylenes can be interpreted in terms of a reduced complex formation under the influence of electron-withdrawing groups, resulting in a relatively smaller contribution of pathway 2 and in a lowering of the rate constant. It is

Table VI. Solvent Dependence of k_{rel} at 0°C

	Olefin			
Solvent	cis-*4-Methyl-2-pentene*	trans-*4-Methyl-2-pentene*	*3-Phenyl-propene*	p-*Methyl-styrene*
CCl_4	2.59	1.87	0.113	1.50
CH_3OH	3.08	2.01	0.147	1.49
$C_2H_4Cl_2$	2.45	1.79	0.123	1.49
$C_2H_5NO_2$	2.75	1.87	0.119	1.51

also noted that with electron-withdrawing substituents the Taft slope ρ^* approaches zero (Figure 5).

The particular behavior of styrene, within the series of monosubstituted ethylenes, can be interpreted in terms of the important conjugation in this olefin. This conjugation increases the polarizability of the reacting π bond, as indicated by the important increase of the molecular refraction which is observed for styrene, compared with an isolated double bond ($\Delta[R]_\alpha = 1.20$ cc/mole) (*24*). As pointed out by Huisgen (*12*), this exaltation of the polarizability favors the concerted 1,3-dipolar cycloaddition. The nucleophilic reaction step is no longer the rate-determining one, as indicated by the negative Hammett ρ-value observed for the ring-substituted styrenes (*13*). Hence, although the two σ-bonds are assumed to be formed "almost" simultaneously in the cycloaddition, their rates of formation differ in that the ozone appears to attack preferentially *via* its positive pole. This increase of the reaction rate also occurs with *trans*-stilbene ($k_{rel} = 3.13$), where the second phenyl group increases the polarizability further with respect to styrene. In *cis*-stilbene, however, where the conjugation is reduced because of the non-planarity of the system, the polarizability is low, with the consequent decrease in the rate constant ($k_{rel} = 0.233$).

The dependence of the relative reaction rates on olefin geometry can be discussed with reference to Equation 14. As pointed out by Murray and co-workers (*25*), the complex formation occurs with cis-olefins rather than with their trans isomers for steric reasons; hence, K_c (cis) > K_c (trans). However, during the formation of the primary ozonide by either path the olefinic carbon atoms change in hybridization, from sp^2 to sp^3. The bond angles thus decrease from 120° to 109°; in the cis isomers, this results in a compression of the substituents van der Waals radii. The repulsion between the substituents is increased, and so is the activation energy. Consequently, k_1 (trans) > k_1 (cis) and k_2 (trans) > k_2 (cis). In the final analysis, the geometry of the olefin has opposite effects (a) on K_c and (b) on k_1 and k_2. Present results seem to indicate that for large substituents the effect on K_c predominates since k (cis) > k (trans).

Finally, the activated complex in path 1 and the π (or σ) complex in path 2 have similar charge separations, which are similar to that of the ozone molecule. Any solvent effect is thus expected to be similar in both mechanisms. This is confirmed by the results in Table VI, which indicate that k_{rel} is nearly independent of the ozonolysis solvent.

Acknowledgment

We thank R. Criegee for preparing and supplying *cis*-di-*tert*-butylethylene. The financial support given by the National Research Council of Canada is gratefully acknowledged.

Literature Cited

1. Badger, G. M., Campbell, J. E., Cook, J. W., Raphael, R. A., Scott, A. J., *J. Chem. Soc.* (1950) 2326.
2. Brown, R. D., *Quart. Rev. Chem. Soc.* (1952) **6,** 63.
3. Brown, R. D., *J. Chem. Soc.* (1950) 3249.
4. van Dijk, J., *Rec. Trav. Chim. Pays-Bas* (1948) **67,** 945.
5. Wibaut, J. P., Sixma, F. L. J., Kampschmidt, L. W. F., Boer, H., *Rec. Trav. Chim. Pays-Bas* (1950) **69,** 1355.
6. Wibaut, J. P., Sixma, F. L. J., *Rec. Trav. Chim. Pays-Bas* (1952) **71,** 761.
7. Bailey, P. S., Ashton, J. B., *J. Org. Chem.* (1957) **22,** 98.
8. Nakagawa, T. W., Andrews, L. J., Keefer, R. M., *J. Amer. Chem. Soc.* (1960) **82,** 269.
9. Vrbaski, T., Cvetanović, R. J., *Can. J. Chem.* (1960) **38,** 1053.
10. Williamson, D. G., Cvetanović, R. J., *J. Amer. Chem. Soc.* (1968) **90,** 3668.
11. *Ibid.*, p. 4248.
12. Huisgen, R., *Angew. Chem., Intern. Ed. Engl.* (1963) **2,** 565, 633.
13. Klutsch, G., Fliszár, S., *Can. J. Chem.*, in press.
14. Taft, R. W., *J. Amer. Chem. Soc.* (1953) **75,** 4231.
15. Taft, R. W., "Steric Effects in Organic Chemistry," M. S. Newman, Ed., Wiley, New York, 1956.
16. Pritzkow, W., Schoppe, G., *J. Prakt. Chem.* (1969) **311,** 689.
17. Bailey, P. S., *Chem. Ind.* (1957) 1148.
18. Bailey, P. S., *Chem. Rev.* (1958) **58,** 925.
19. Bailey, P. S., Lane, A. G., *J. Amer. Chem. Soc.* (1967) **89,** 4473.
20. Murray, R. W., Youssefyeh, R. D., Story, P. R., *J. Amer. Chem. Soc.* (1967) **89,** 2429.
21. Durham, L. J., Greenwood, F. L., *Chem. Commun.* (1967) 843.
22. Durham, L. J., Greenwood, F. L., *J. Org. Chem.* (1968) **33,** 1629.
23. Bailey, P. S., Potts, F. E., Ward, J. W., *J. Amer. Chem. Soc.* (1970) **92,** 230.
24. Ingold, C. K., in "Structure and Mechanisms," G. Bell, London, 1953.
25. Murray, R. W., Youssefyeh, R. D., Story, P. R., *J. Amer. Chem. Soc.* (1966) **88,** 3143.

RECEIVED May 20, 1971.

5

The Interaction of Ozone with Double Bonds Containing Vinyl Bromide Moieties

KARL GRIESBAUM and J. BRÜGGEMANN

Engler-Bunte-Institut der Universität Karlsruhe, 75 Karlsruhe, Germany

The present contribution reports the first results of a systematic approach to a study of the reaction and the products obtained in the liquid-phase ozonolysis of unsaturated substrates containing halogenated double bonds. As a model case the ozonolysis of trans-*2,3-dibromo-2-butene was studied. Ozonolysis occurs at a slower rate than that of pure hydrocarbon olefins and less than equimolar amounts of ozone are sufficient for the quantitative conversion of the substrate. Double bond cleavage is not the overriding reaction, but the original double bond is to a large extent converted into C—C single bonds which are fairly stable towards further ozone attack.*

The ozonolysis of pure hydrocarbon olefins has been studied by generations of chemists ever since 1905 when Harries established the interaction of ozone with double bonds. During this continued research practically all aspects of the ozonolysis reaction have been extensively scrutinized, starting from the nature and products of the initial ozone–double bond interaction, *via* the mechanistic and stereochemical course of the ozone cleavage, to the correlation between the nature of the starting material and the reaction products. Consequently, the ozonolysis of hydrocarbon olefins is rather well understood.

In contrast to this, very little is known about the ozonolysis of olefins which bear halogen substituents at the double bond. This is somewhat surprising since compounds containing vinyl halide moieties are important technical products whose properties could be adversely affected by ozone degradation. A case in point is neoprene rubber, whose performance as an elastomer could suffer considerably by ozone attack and concurrent crack formation (*1*).

Background

When this work was initiated, there were only a few scattered reports in the literature concerned with the interaction of ozone and halogenated double bonds. The bulk of these reports dealt with the ozonolysis of substrates containing both halogenated and non-halogenated double bonds. In all of these cases the non-halogenated double bond was attacked preferentially by ozone, and only products derived from that kind of reaction were described.

Ozonolysis of 1,2,3,4,7,7-hexachlorobicyclo[2.2.1]hepta-2,5-diene, **1**, in participating and in non-participating solvents was reported to yield the cleavage products, **2** and **3**, respectively (*2*):

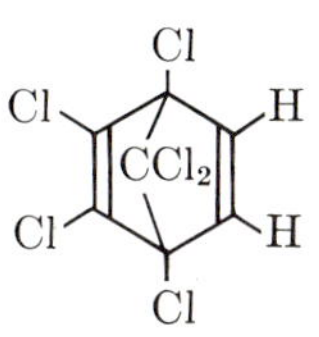

1

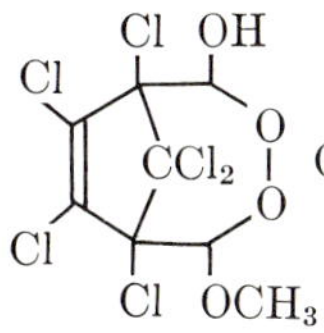

2

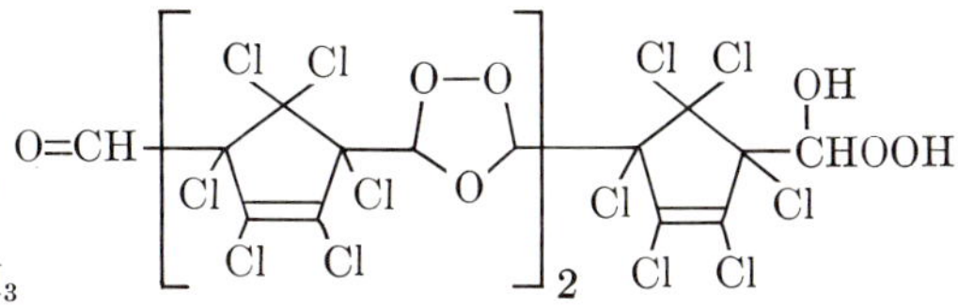

3

Ozonolysis of 1,6,7,8-tetrachloro-2,3,4,5-tetramethylbicyclo[4.2.0.0.(2,5)]octadiene, **4**, in pentane was reported to form the monomeric ozonide, **5** (*3*):

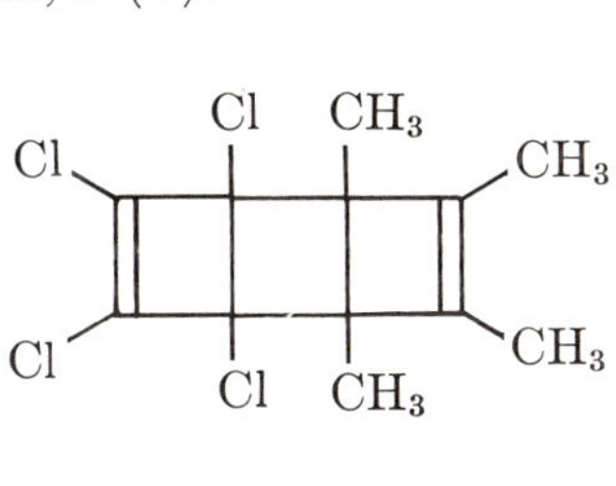

4

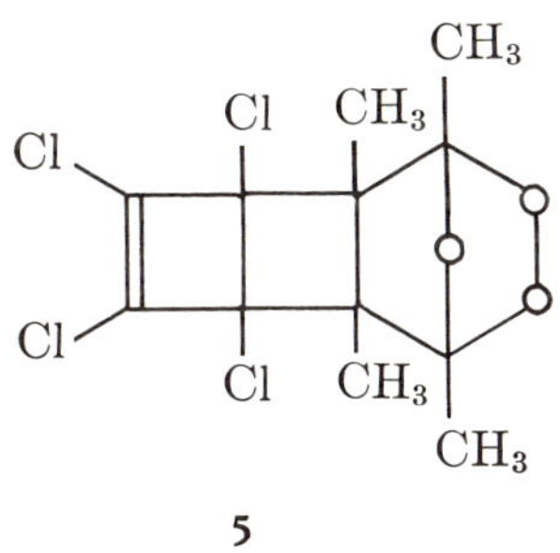

5

Ozonolysis of hexachlorodicyclopentadiene, **6**, in methanol–methylene chloride solution produced **7** as a major ozonolysis product (*4*):

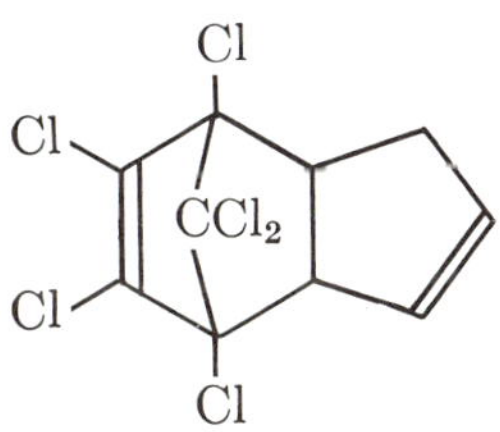

6

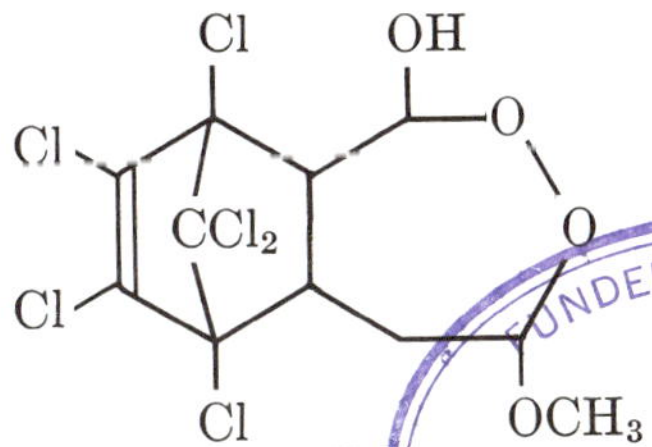

7

Ozonolysis of the monobromide **8** in ethyl acetate occurred also preferably at the non-halogenated double bond to form the ozonide **9** (*5*):

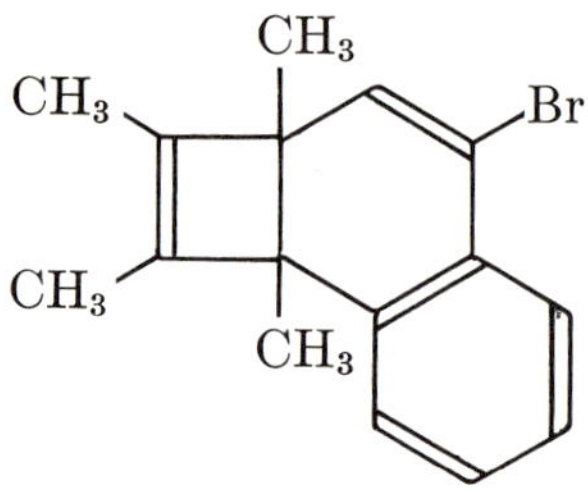

8

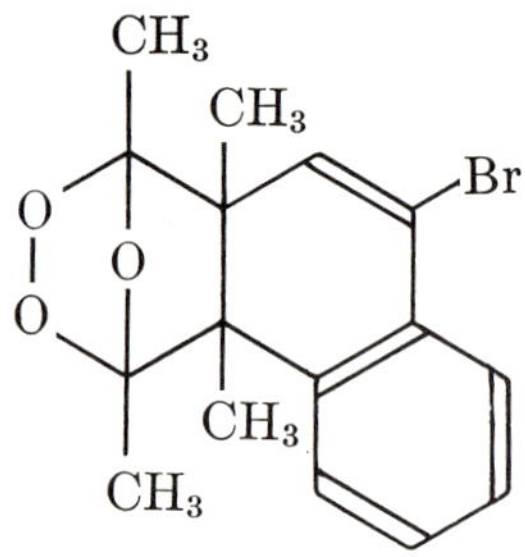

9

These scattered observations indicate that ozonolysis of a halogenated double bond occurs at a much slower rate than that of a non-halogenated double bond, even if the former is part of the more strained ring system as in **6** or if the double is monohalogenated as in **8**. This qualitative picture was substantially confirmed by the results of a quantitative study of the rate of ozone attack at variously chlorinated ethylenes (*6*). As shown in Table I, the rates of ozone attack decrease dramatically as chlorine is successively substituted for hydrogen in ethylene.

Table I. Second-Order Rate Constants (*K*) for the Reactions of Ozone with Olefins in CCl_4 Solutions at Room Temperature (*6*)

Olefin	K, *liter/mole sec*
$CCl_2{=}CCl_2$	1.0
$CHCl{=}CCl_2$	3.6
$CH_2{=}CCl_2$	22.1
cis-CHCl=CHCl	35.7
trans-CHCl=CHCl	591
$CH_2{=}CHCl$	1118
$CH_2{=}CH_2$	25,000 (estimated)

This decreased reactivity of halogenated double bonds towards ozone has been ascribed by some workers to the inductive electron-withdrawing effect of the halogen substituents and to a concurrent decrease of the electron density at the double bond (*6*); others argued more on the basis of steric hindrance of the double bond by the bulky halogen substituents (*4*). In any event, the slow reaction is probably the major reason for the fact that little is known about the products and thus about the course of the ozonolysis of halogenated double bonds. In fact, to our knowledge the reaction products have only been thoroughly examined for the ozonolysis of certain special substrates such as tetrafluoroethylene (*7*) and the 9,10-dihaloanthracenes (*8*).

Ozonolysis of tetrafluoroethylene **10** in non-participating solvents was reported to produce carbonyl fluoride **11** and tetrafluoroethylene oxide **13** as the major compounds, along with minor amounts of perfluorocyclopropane **15** as well as trace amounts of a compound to which the authors assigned the structure of the monomeric ozonide **16** of tetrafluoroethylene (*7*). This result was rationalized by assuming a normal Criegee-type cleavage *via* the two fragments **11** and **12**, followed by secondary reactions of the zwitterion **12** such as epoxidation of tetrafluoroethylene and deoxygenation of **12** to form the carbene **14**.

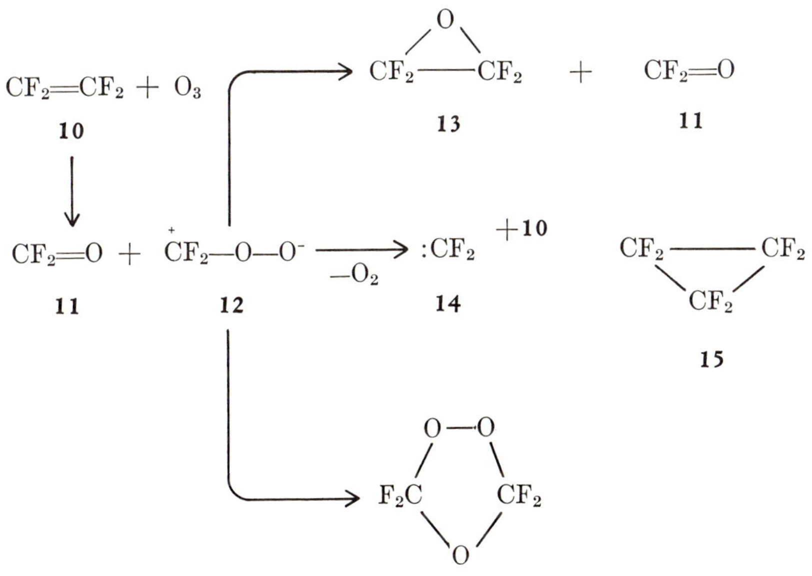

Ozonolysis of 9,10-dibromo- (**17a**) and 9,10-dichloroanthracene (**17b**) was reported to occur partly by attack at the non-halogenated double bonds to yield the di- (**18**) and tetracarboxylic acids, **19**, as well as by attack at the 9,10-positions to yield the dehalogenated product anthraquinone, **22** (*8*). This dual attack might at first seem to be in contrast to what was said above about the relative rates of ozone attack at halogenated and non-halogenated double bonds. However, this is not the case if one considers that (in analogy to the ozonolysis of the unsubstituted anthracene) the reaction at the 9,10-positions has to be formulated as an atom- rather than a bond attack. In accordance with such a rationalization, the authors (*8*) formulated the intermediates **20** and **21** as precursors for anthraquinone.

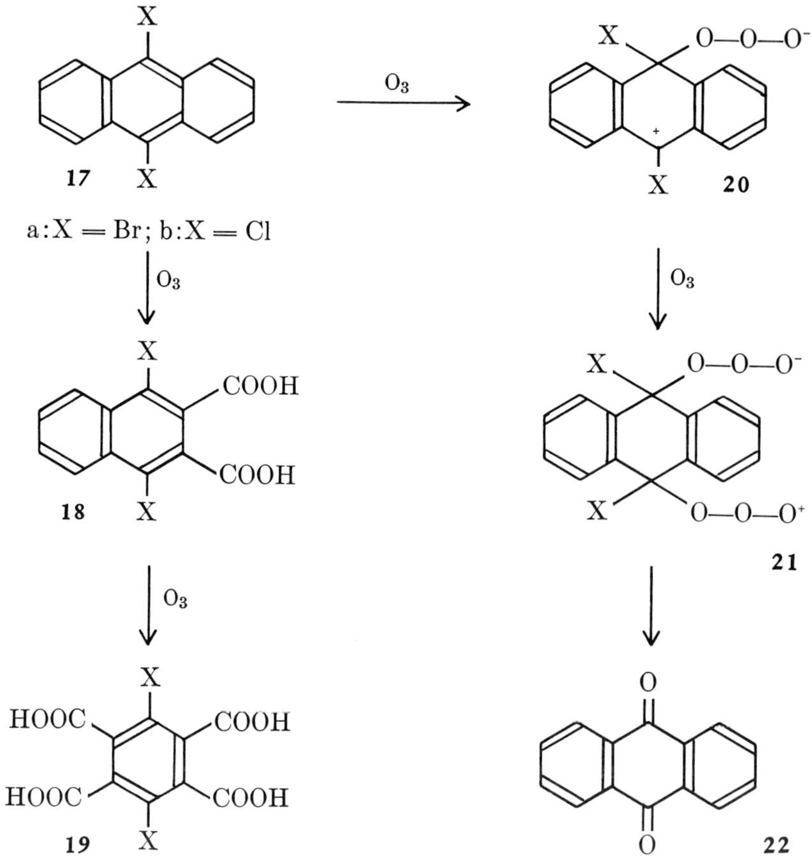

It is apparent from the foregoing that these special cases cannot be viewed as typical examples for the ozonolysis of halogenated double bonds in general. In the ozonolysis of tetrafluoroethylene there was partial cleavage of the double bond and no loss of halogen substituents, while ozonolysis of the 9,10-dihaloanthracenes resulted in the loss of the halogen substituents, however, not in a cleavage of the halogenated double bonds. To obtain information of a more general nature, it appeared necessary to select model substrates which more closely resemble the structures of the commonly occurring mono- and dihalosubstituted olefins. As the first model we chose 2,3-dibromo-2-butene, whose ozonolysis is described below.

Experimental

Ozonolysis of 2,3-Dibromo-2-butene. The ozone–oxygen stream that left the ozonizer passed through a trap which was kept at −78°C in order to keep the gas free of moisture. The gas then passed through the reaction mixture and then to a condenser kept at −78°C to minimize losses of low boiling components. Before entering an aqueous sodium iodide solution, the exit gas again passed a cold trap to avoid condensation of moisture from the sodium iodide solution into the reaction mixture. Samples for NMR analysis were taken by transferring the reaction mixture pneumatically into a previously attached NMR tube so that during this operation also admission of moisture was minimized.

Ozonolysis of *trans*-2,3-Dibromo-2-butene in Inert Solvents. STARTING MATERIAL. The title compound **23** was prepared by adding bromine to 2-butyne at *ca.* −30° to −40°C. The reaction produced **23** in more than 95% yield, along with less than 5% of the cis-isomer, **24**, and trace amounts of as yet unidentified by-products.

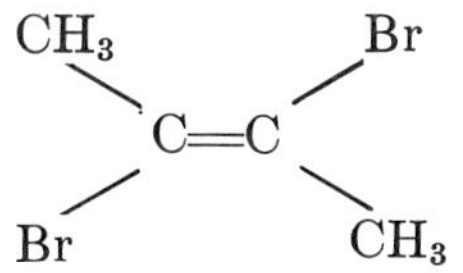

23

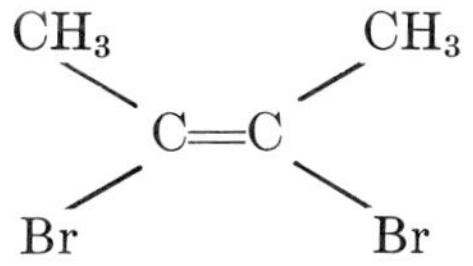

24

Results

The stereochemical assignments of **23** and **24** are based on the infrared spectra of the isolated pure compounds (Figure 1). The spectrum of the cis-isomer, **24**, shows a sharp, presumably C=C stretching band at 1640 cm^{-1} which is absent in the spectrum of the trans-isomer, **23**.

Stoichiometry of the Reaction. Upon ozonolysis of **23** in inert solvents (pentane, methylene chloride, 1,1,2,2-tetrachloroethane) or without solvents, the originally colorless solutions turned light brown within minutes after the ozonolysis began and gradually assumed the typical deep red color of dissolved bromine. Unreacted ozone began to pass the reactor a few minutes after the ozonolysis began, and at any time thereafter the amount of ozone produced surpassed the amount of ozone consumed (Figure 2). This result was undoubtedly caused by the decreased reactivity of the dibrominated double bond in **23**.

To assess the quantitative correlations between ozone consumption, olefin consumption and product formation, experiments were carried out in the presence of 1,1,2,2-tetrachloroethane as an internal standard for NMR analyses. These experiments showed that only about half the equimolar amount of ozone was required to consume the olefin completely (Figure 3). This stoichiometry was obviously not the 1:1-stoichiometry usually found in the ozonolysis of hydrocarbon olefins and which is the basis for the quantitative assessment of the number of double bonds in organic molecules by ozonolysis. Attempts to explain this unusual stoi-

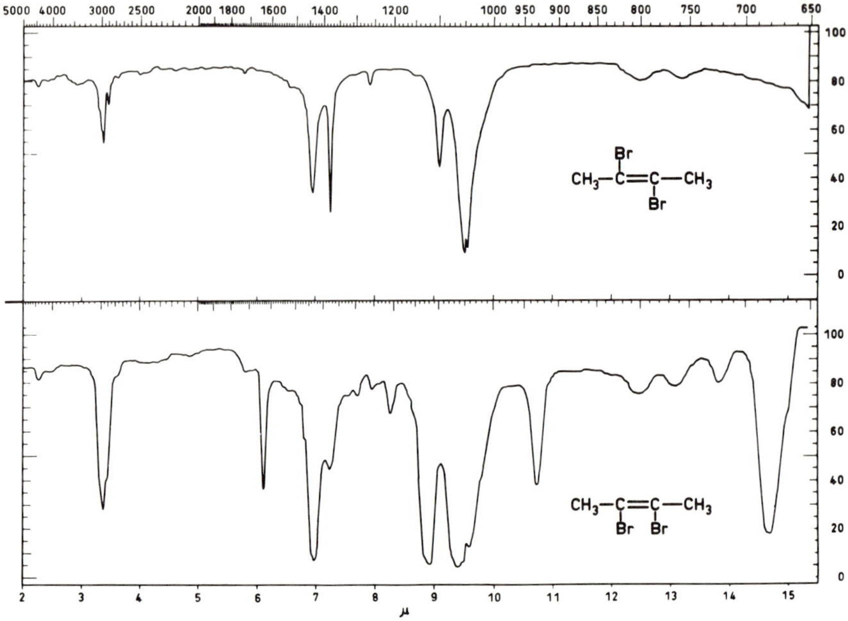

Figure 1. Infrared spectra of cis- *and* trans-*2,3-dibromo-2-butene (neat)*

chiometry had to await the qualitative and quantitative analysis of the reaction products.

Ozonolysis Products. Preliminary experiments showed that the ozonolysis of **23** produced a number of highly reactive components such as bromine, acetyl bromide, and acetic acid. Therefore, a quantitative analysis of the crude reaction products by GLC was not feasible. NMR analysis, on the other hand, was easily possible since both the starting material and most of the products exhibited singlet signals that were well enough separated for quantitative analysis. Figure 4 shows the NMR spectrum of a crude ozonolysis product in which the substrate **23** had been completely converted.

Each structural assignment of signals 1–7 in the spectrum of Figure 4 is based on the addition of the authentic compound to the crude ozonolysis mixture, and—aside from peak 1—is also based on additional analytical evidence: 2,2,3,3-tetrabromobutane (peak 2), 3,3-dibromobutanone (peak 4), and acetic acid were actually isolated while acetyl bromide (peak 3) as well as acetic anhydride (peak 5) were further identified by their sensitivity to solvolysis reactions, particularly hydrolysis and alcoholysis to form the corresponding esters. Diacetylperoxide (peak 6) was identified by the disappearance of peak 6 from the NMR spectrum as well as the disappearance of the typical infrared bands at 1810 and 1835 cm^{-1} when the reaction mixture was treated with sodium iodide.

In addition to the seven organic products mentioned above, ozonolysis of **23** also produced considerable amounts of free bromine, which was qualitatively and quantitatively detected by its addition to 2,3-dimethyl-2-butene and subsequent analysis of the resulting dibromide.

Discussion

Rationalization of Ozonolysis Products. Obviously the ozonolysis of *trans*-2,3-dibromo-2-butene leads to a greater variety of products than one might have expected if the reaction were strictly analogous to the ozonolysis of hydrocarbon olefins. In particular, there is no evidence for the formation of the trioxolane-type ozonide **27**. Nevertheless, all the reaction products identified could be readily rationalized by a normal Criegee-type primary cleavage pattern if one assumes that, owing to their special structural features, the two fragments, acetyl bromide **25** and the zwitterion **26**, undergo a number of subsequent reactions which are not open to the ozonolysis fragments derived from purely hydrocarbon olefins:

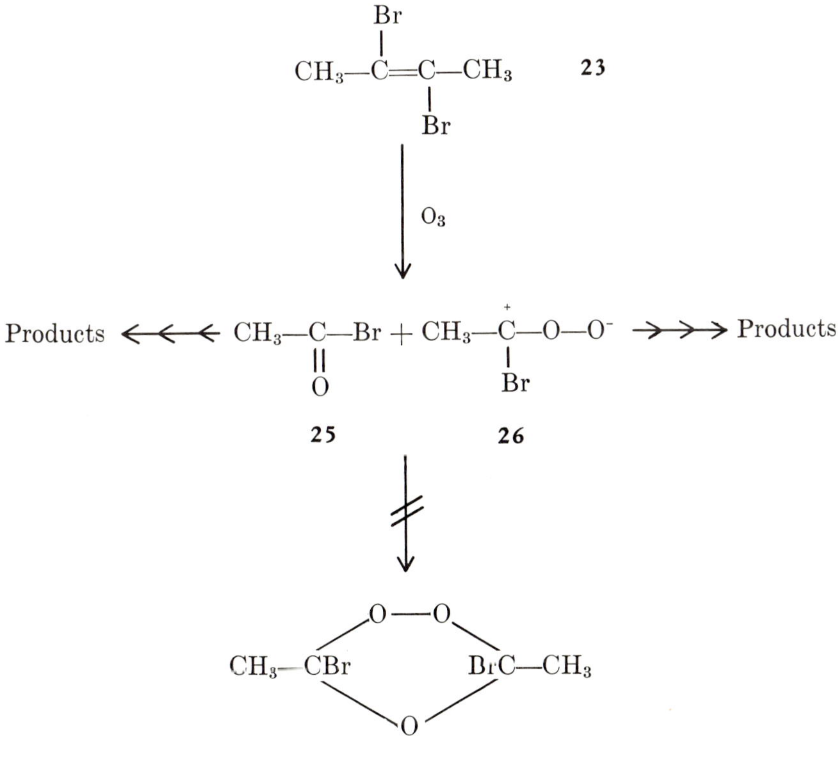

23

25 **26**

27

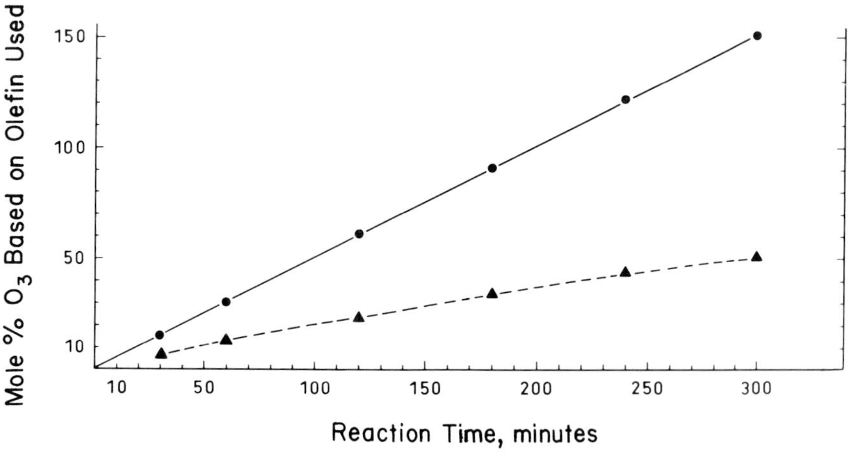

Figure 2. Ozone produced (●) vs. *ozone consumed* (▲)

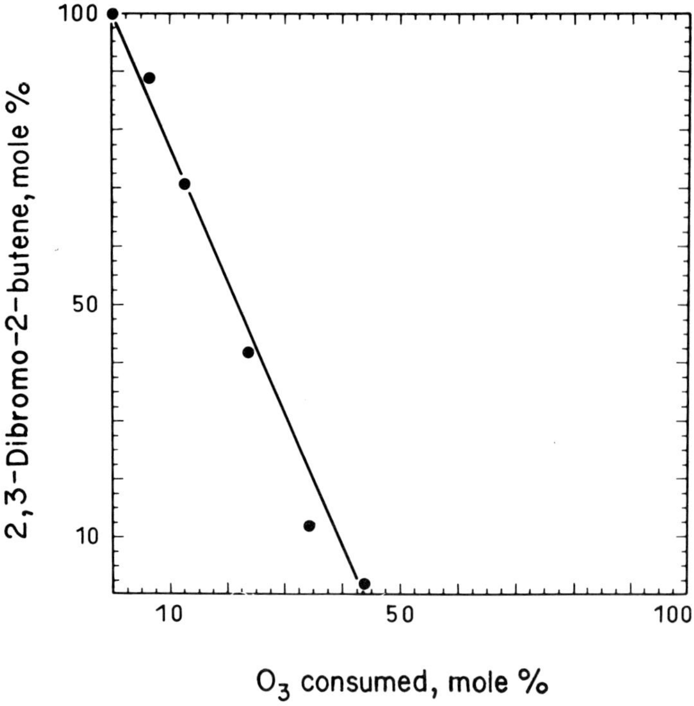

Figure 3. Ozone consumption vs. *olefin consumption*

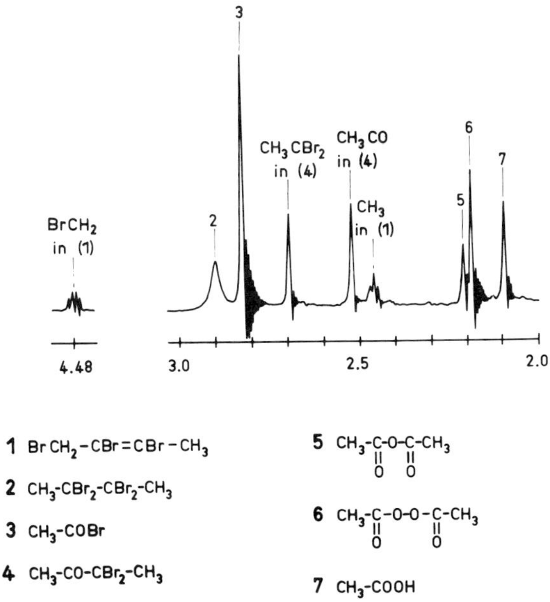

Figure 4. NMR spectrum of the ozonolysis product of trans-*2,3-dibromo-2-butene*

The acetyl bromide fragment, **25**, is most likely the precursor for acetic acid, **28**, acetic anhydride, **29**, and for part of the bromine that is formed during the ozonolysis reaction. Although the reactions were carried out under anhydrous conditions, adventitious amounts of moisture could initiate the following reaction sequence:

$$\underset{\mathbf{25}}{CH_3COBr} + H_2O \rightarrow \underset{\mathbf{28}}{CH_3COOH} + HBr$$

$$CH_3COBr + CH_3COOH \rightarrow \underset{\mathbf{29}}{CH_3CO{-}O{-}COCH_3} + HBr$$

$$2HBr + O_3 \rightarrow H_2O + Br_2 + O_2$$

Spontaneous hydrolysis of acetyl bromide, **25**, leads to the formation of acetic acid, **28**, acetic anhydride, **29**, and hydrogen bromide. Since there is always excess ozone present, hydrogen bromide is rapidly reoxidized to form bromine and to reform the water. Therefore, the sequence depicted above can be summarily represented by the following reaction:

$$2CH_3COBr + O_3 \rightarrow CH_3CO{-}O{-}COCH_3 + Br_2 + O_2$$

Such a reaction course has indeed been verified by independent experiments. Upon treatment of acetyl bromide with ozone under conditions that were comparable with those used in the ozonolysis of the dibromoolefin **23**, the reaction produced bromine and acetic anhydride in more than 90% yield.

The zwitterion fragment, **26**, is probably the precursor for the formation of 3,3-dibromobutanone, **31** and of diacetyl peroxide, **33**. The former compound, **31**, is probably formed through the epoxidation of the starting material, **23**, by the zwitterion, **26**, and subsequent spontaneous rearrangement of the dibromoepoxide, **30**. Such a reaction sequence has been rendered likely by the results of the direct epoxidation of **23** with *m*-chloroperbenzoic acid which also produced the rearrangement product, **31**.

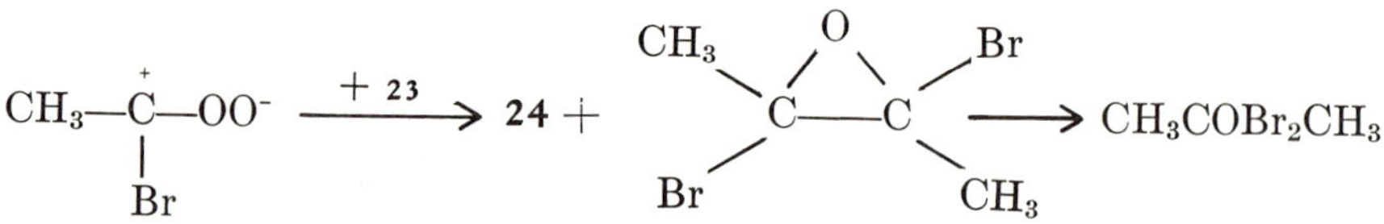

26 **30** **31**

The formation of the completely debrominated product, diacetyl peroxide **33**, from the zwitterion intermediate, **26**, probably proceeds *via* the dimeric peroxide of structure **32** in the manner depicted below:

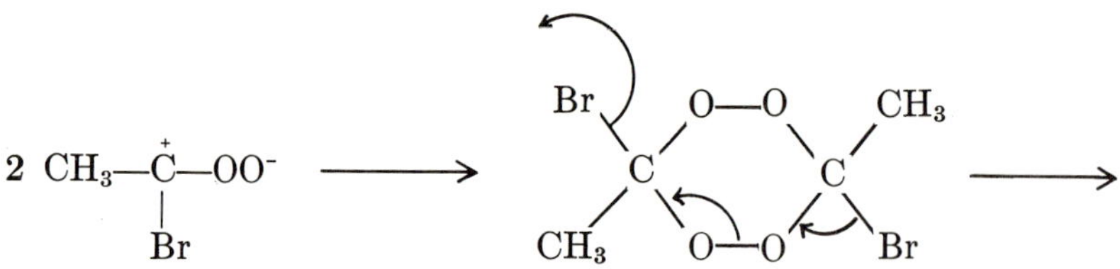

32

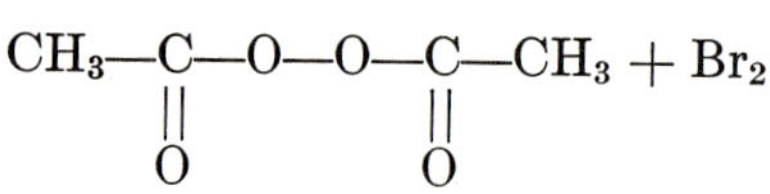

33

Although the intermediate dimeric peroxide, **32**, could not be isolated and rigorously proved, there are two items of evidence which favor the postulated reaction course. The first evidence comes from independent experiments in which tetracyanoethylene was used as a coreagent during the ozonolysis of **23**. In these cases, the formation of diacetyl peroxide, **33**, as well as that of 3,3-dibromobutanone, **31**, was drastically reduced, and at the same time tetracyanoethylene epoxide was formed. Apparently, the reaction of tetracyanoethylene with the zwitterion, **26**, is faster than the dimerization of the zwitterion to form **32**—a phenomenon already observed during the ozonolysis of normal olefins (9).

The second piece of evidence for the intermediacy of the dimeric peroxide, **32**, stems from the low temperature ozonolysis of *trans*-2,3-dibromo-2-butene, **23**. In contrast to the usual reactions which were carried out at *ca.* −35°C, the ozonolysis of **23** in pentane at −78°C yielded a solid white precipitate which exploded violently when we tried to isolate it. When the low temperature ozonolysis was carried out in methylene chloride, on the other hand, no precipitate appeared. However, when the methylene chloride solution of the ozonolysis product was allowed to warm up gradually, an exothermic reaction set in at around −50°C, and the originally colorless solution showed the typical color of dissolved bromine. It is assumed that the explosive material is the dimeric peroxide, **32**, which undergoes spontaneous debromination to form diacetyl peroxide, **33**.

The remaining two products from the ozonolysis of **23**—*viz.*, the tetrabromide, **34**, and the tribromide, **35**, are the result of addition and substitution reactions of the bromine that is liberated during the ozonolysis reaction. Both reactions are probably favored by the presence of peroxidic materials in the solution since it could be shown that they progress much slower if bromine and *trans*-2,3-dibromo-2-butene, **23**, react in the absence of initiators. This is particularly true for the substitution reaction leading to **35**. During independent experiments, when the substrate, **23**, was added to ozonized solutions of bromine in methylene chloride, the substitution was markedly favored.

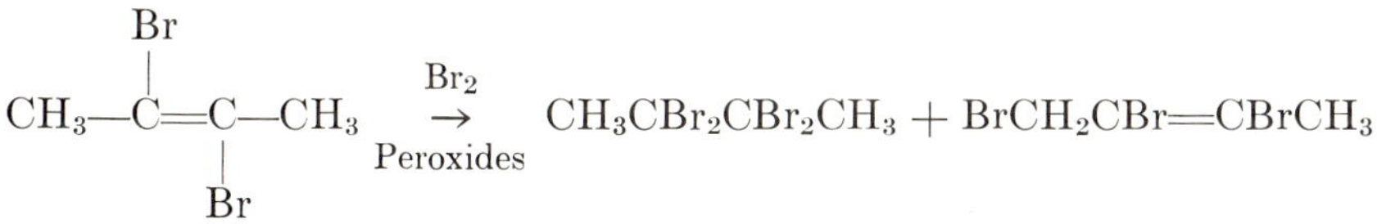

23 **34** **35**

The nature and the distribution (Table II) of the ozonolysis products in conjunction with the probable modes of their formation allow also a qualitative rationalization of the observed ozone–olefin stoichiometry. Three reactions compete with ozone for the starting material, *trans*-2,3-dibromo-2-butene. These reactions are the formation of 3,3-dibromobutanone, **31**, and the formation of the brominated products, **34** and **35**. On the other hand, hydrogen bromide is oxidized to form bromine and water, which consumes ozone on top of the regular olefin–ozonolysis reaction. An attempt to explain the observed stoichiometry quantitatively did, however, not lead to a satisfactory correlation between the actual ozone consumption and the observed material balance. This

Table II. Products of the Ozonolysis

Solvent	*Conversion, %*	*Product Distribution, mole %*			
		AcBr	*AcOH*	*AcOAc*	*AcOOAc*
none	50	26	15	8	9
Pentane	85	44	19	—	6
CH_2Cl_2	95	23	24	14	13

could be the result of a concurrently occurring, ozone-initiated autoxidation reaction such as ozone-initiated oxidation of hydrogen bromide.

Ozonolysis of *trans*-2,3-Dibromo-2-butene in Participating Solvents. In contrast to the great variety of products resulting from the ozonolysis of **23** in inert solvents, the ozonolysis in methanol, ethanol, and 1-butanol produced mainly the corresponding acetates, **37**, and bromine. In addition, the presence of an as yet unidentified peroxidic material was detected by the sodium iodide method. However, in each case the amount of ester produced was greater than one would have expected if the acetyl bromide fragment, **25**, were the only precursor for it. This seems to indicate that the second, presumably zwitterion fragment, **26**, can also lead to the formation of ester. Obviously, this would require the reduction of a peroxidic intermediate, possibly of the type **36**. This may proceed in a manner similar to that formulated for the debromination of the dimeric peroxide, **32**. The hypobromite that is also formed during this step can subsequently oxidize the hydrogen bromide produced during the alcoholysis of acetyl bromide.

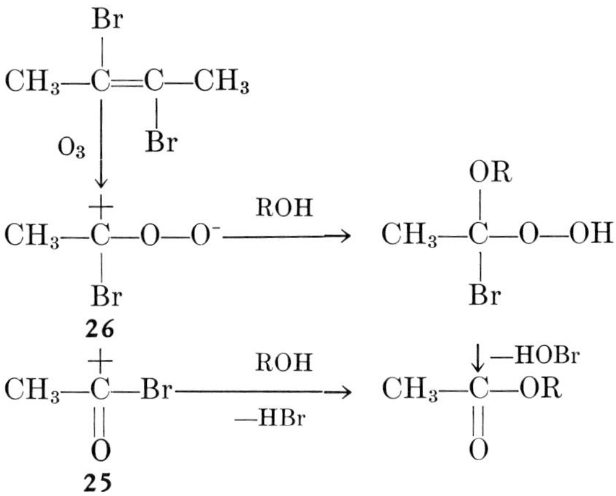

36

37

Conclusions

If one disregards the tribromide, **35**, in which the double bond of the substrate, **23**, has not been touched at all, the ozonolysis of *trans-*

of 2,3-Dibromo-2-butene

Product Distribution, mole %		
$CH_3COCBr_2CH_3$	$CH_3CBr_2CBr_2CH_3$	$CH_3CBr{=}CBrCH_2Br$
14	16	11
20	6	5
13	9	4

2,3-dibromo-2-butene, **23**, in inert solvents leads essentially to three types of reaction products—the two actual cleavage products, **25** and **28**, two products (**29** and **33**) in which the carbon atoms of the original double bond are still held together by a labile anhydride and peroxide group, respectively, and two products (**31** and **34**) in which the original double bond has been converted to single bonds that are stable towards further ozone attack:

$$\underset{\mathbf{23}}{CH_3{-}CBr{=}CBr{-}CH_3} + O_3 \rightarrow \underset{\mathbf{25}}{CH_3COBr} + \underset{\mathbf{28}}{CH_3COOH}$$

$$\rightarrow \underset{\mathbf{29}}{CH_3CO{-}O{-}COCH_3} + \underset{\mathbf{33}}{CH_3CO{-}OO{-}COCH_3}$$

$$\rightarrow \underset{\mathbf{31}}{CH_3COCBr_2CH_3} + \underset{\mathbf{34}}{CH_3CBr_2CBr_2CH_3}$$

Although the distribution among these products varies, depending on the reaction conditions, particularly on the solvent used, it is evident from the data in Table II that the latter type of products in which the double bond has been immunized towards further ozone attack comprises in each case a significant part (between 22 and 30%) of the total product mixture. This is perhaps the most remarkable result of the present investigation. In contrast to the ozonolysis of hydrocarbon olefins, the ozonolysis of this dibromosubstituted double bond cannot be viewed primarily as a double bond cleavage reaction.

This result may also give a partial explanation for the known ozone stability of polymers containing vinyl halide moieties. Part of this inertness is undoubtedly the result of the slow ozone attack. However, this phenomenon can only impart a prolonged ozone resistance, while the conversion of the original double bond into single bonds could impart a permanent ozone resistance. We are extending our research to further model compounds of such polymer microstructures to find out whether this is a general phenomenon of the ozonolysis of halogenated double bonds.

From a preparative standpoint, ozonolysis in inert solvents does not represent a satisfactory method for the ozone cleavage of dibromo-substituted double bonds. If this is the desired reaction, it is preferable to use alcohols as participating solvents, which lead to a clean cleavage and to a high yield of the corresponding ester fragments. The latter reaction constitutes a convenient preparative method for the cleavage of acetylenes into the corresponding ester fragments *via* the intermediate dibromination products.

Literature Cited

1. Kondo, A., *Raba Daijesuto (Japan)* (1968) **20** (3), 42.
2. Slagel, R. C., *J. Org. Chem.* (1966) **31,** 593.
3. Criegee, R., Huber, H., *Angew. Chem.* (1969) **81,** 749.
4. Franz, J. E., Knowles, W. S., Osuch, C. J., *J. Org. Chem.* (1965) **30,** 4328.
5. Criegee, R., Schweickhardt, C., Knoche, H., *Chem. Ber.* (1970) **103,** 960.
6. Williamson, D. G., Cvetanović, R. J., *J. Amer. Chem. Soc.* (1968) **90,** 4248.
7. Gozzo, F., Camaggi, G., *Chim. Ind. (Milan)* (1968) **50,** 197.
8. Kolsaker, P., Bailey, P. S., Dobinson, F., Kumar, B., *J. Org. Chem.* (1964) **29,** 1409.
9. Criegee, R., Günther, P., *Ber.* (1963) **96,** 1564.

RECEIVED August 23, 1971.

6

The Ozone–Hydrosilane Reaction: A Mechanistic Study

LEONARD SPIALTER, LeROY PAZDERNIK, STANLEY BERNSTEIN, WILLIAM A. SWANSIGER, GLEN R. BUELL, and MICHAEL E. FREEBURGER

Chemical Research Laboratory, Aerospace Research Laboratories, Wright-Patterson AFB, Ohio 45433

The reaction of a hydrosilane with ozone results in the rapid, quantitative conversion of the Si–H bond to the Si–OH moiety. The mechanism of this conversion has now been elucidated. It involves a fast, reversible complexation of ozone (acting as a nucleophile) with the silicon atom, followed by rate-determining electrophilic attack by the bound ozone upon the hydridic hydrogen, and decomposition into a $R_3Si\cdot\ \cdot OH$ *radical pair which recombine to produce the silanol. Extensive data concerning the relative rates and other structure-dependent properties in the ozonation of a number of mono-, di-, and trihydrosilanes are presented.*

The use of ozone as an oxidant for organic substrates has been known for several decades, and the scope and mechanism(s) of these oxidations have experienced considerable investigation (*1, 2*). However, it was not until 1962 that the initial report of the ozone oxidation (ozonation) of organosilanes appeared (*3*). In 1965, it was discovered that ozone would rapidly, cleanly, and quantitatively convert an Si–H moiety to the corresponding silanol (Si–OH) (*4*). Subsequent investigations of this reaction elucidated its scope and resulted in suggested mechanistic pathways (*5–11*). Although the ozonation of the C–H bond is relatively slow and yields, in general, a mixture of products (*1, 12–15*), that of the Si–H bond is rapid and affords only silanol. The present paper describes a rational mechanism for the ozonation of the Si–H bond. During the investigation, considerable data concerning substituent electronic effects on silicon were accumulated and are described. Their implications in organosilicon chemistry are also discussed.

Experimental

Materials. Solvents used included hexane, carbon tetrachloride, and methylene chloride, all of spectroquality and further purified by saturation with ozone followed by nitrogen purging and distillation. For ozone generation, oxygen gas, dried by passage through a dry ice-acetone cold trap, was introduced into a Welsbach model T-23 laboratory ozonator operated at 10 volts and 7 psig. The effluent gas, about 4% ozone in oxygen, was passed through an adjustable stream splitter for flow control and then bubbled through a saturator containing the solvent also being used in the reaction system. (For the experiments with oxygen-free ozone, the ozone was first selectively adsorbed from the ozonator effluent stream onto silica gel at −77°C and then desorbed by warming and elution by a dry argon stream.)

The hydrosilane reagents used, where available, were obtained from commercial suppliers such as Pierce Chemical (Rockford, Ill.), Peninsular Chem Research (Gainesville, Fla.), and Matheson, Coleman and Bell (Norwood, Ohio). Others were synthesized by the lithium aluminum hydride or deuteride (Alfa Chemical, Beverly, Mass.) reduction of appropriate organic chloro- or fluorosilane, either purchased from the above-mentioned sources or synthesized by conventional routes involving organomagnesium or -lithium condensations with halosilanes. All such compounds were at least 98% pure as determined by vapor-phase chromatography(VPC). Structural identity was established by molecular analysis, infrared, and NMR spectroscopy where necessary.

A new compound synthesized was 2,2′-biphenylenylsilane, by $LiAlH_4$ reduction of 2,2-biphenylenyldifluorosilane (provided by Anderson Chemical Division, Stauffer Chemical Co., Adrian, Mich.) with m.p. 36.6°C, b.p. 86°C (0.31 mm.). *Anal.* Calcd for $SiC_{12}H_{10}$:Si, 15.41; C, 79.06; H, 5.53. Found (Galbraith Microanalytical Lab, Knoxville, Tenn.): Si 15.12; C, 78.92; H, 5.61.

Ozone Competition Reaction Procedures. In the relative rate studies, the solvent-saturated ozone–oxygen stream was passed into a glass bubbler reactor vessel charged with 4 ml of about $4 \times 10^{-2}M$ concentration of each of the two silanes to be competitively ozonized as well as of an inert saturated aliphatic hydrocarbon to function as internal standard. (For example, *n*-undecane was used for the tributylsilane/trihexylsilane study.) The effluent from the reactor passed through a solvent-filled bubble counter to visualize the flow. The inlet stream splitter mentioned earlier was adjusted to allow 2–4 hours for each run's completion, as determined by experience. The temperature was controlled at 0°C in both the saturator and the reactor by an ice bath.

To follow the course of the reaction, small (0.1 ml) aliquots were periodically removed and analyzed by gas chromatography on a 150-ft 0.1 inch id capillary column coated with Dow-Corning DC-550 silicone. Peak areas were determined and converted to concentration values by reference to the area of the internal standard peak. From the unreacted fraction $(Si)_t/(Si)_o$, of each paired competing silane Si at sampling time *t* was computed the relative rate constant k_{rel}, the ratio of the corresponding first order (in silane) rate constants, by the earlier reported (*5, 11*) equation:

Table I. Relative Rate/$\sum^{4}\sigma^*$ Correlation

Silane	k_{rel} [a]	log k_{rel}	$\sum^{4}\sigma^{*b}$	$\sum^{4}\sigma^{*e}$
Tris(perhydro-l-naphthyl)silane [c]	378	2.578	−0.29 [d]	−.29
Tricyclohexylsilane	236	2.373	.04	−.12
tert-Butyldicyclohexylsilane	226	2.354	—.11	−.11
Trihexylsilane	115	2.061	.07	.12
Tributylsilane	100	2.000	.10	.17
Triethylsilane	84	1.924	.19	.23
3,3,3-Trifluoropropyldimethylsilane	51.7	1.714	.48 [d]	.40
Dichloromethyldimethylsilane	41.7	1.620	.48 [d]	.48
Tris(3,3,3-trifluoropropyl)silane	35.6	1.551	.46 [d]	.53
Phenyldimethylsilane	30.7	1.487	.54 [d]	.58
Diisopropylsilane	28.5	1.455	.60	.61
Chloromethyldimethylsilane	23.6	1.373	.67 [d]	.67
Diphenylmethylsilane	23.0	1.362	.59 [d]	.68
Triphenylsilane	23.0	1.362	.64 [d]	.68
Dibutylsilane	21.7	1.336	.72	.70
Triethoxysilane	19.0	1.278	.76 [d]	.75
Phenylmethylsilane	12.0	1.079	1.03 [d]	.91
2,2-Diphenylenylsilane	9.6	0.982	.99 [d]	.99
Diphenylsilane	7.9	.898	1.08 [d]	1.05
Cyclohexylsilane	3.65	.562	1.32	1.32
Dibenzylsilane	3.31	.520	1.42	1.36
Hexylsilane	3.04	.483	1.33	1.38
Phenylsilane	2.10	.322	1.52 [d]	1.51

[a] All rates reduced to 0°C and relative to Bu_3SiH rate of 100.
[b] Where applicable, Taft σ* values for substituents are taken from tabulations in J. Hine, "Physical Organic Chemistry," 2nd ed., p. 97, McGraw-Hill, New York, 1962.
[c] Mixture of 1-decalinyl isomers obtained by lithium aluminum hydride reduction of hydrogenated tris-l-naphthyl-naphthylfluorosilane (*cf.* Ref. *5*.)
[d] Calculated using σ* values of this paper, Table II.
[e] Calculated from the equation: $\Sigma\sigma^* = (2.2166 - \log k_{rel})/1.2513$ of Figure 1.

$$k_{rel} = k_1/k_2 = \{\log[(S_2)_t/(S_2)_o]\}/\{\log[(S_1)_t/(S_1)_o]\}$$

[Note: Equation 4 in Ref. *11* has a typographical error resulting in a numerator–denominator interchange error.] Statistical reduction of data from multiple samples was accomplished by a non-linear least squares computer program (*16*).

The k_{rel} values of this paper are normalized to 100 for tributylsilane rather than to the earlier triethylsilane whose greater volatility introduced evaporative losses and resultant data scatter in the experiments.

Results and Discussion

Relative Reactivity and Electronic Considerations. The initial thrust of this investigation was to determine the relative reactivity of selected organosilanes towards ozone and to correlate the reactivity with the substituents, either structurally or electronically. A linear relationship between the Si–H stretching frequency (~2200 cm^{-1}) of monohydrosilanes

(R_3SiH) and the logarithm of their relative rates of ozonation has been previously reported (*5, 8*). The Si–H stretching frequency has also been correlated with an effective group electronegativity (*17, 18*) and with the Taft inductive parameter, σ^* (*19, 20*); however, certain substituents (chloro, fluoro, methoxy, phenyl, dimethylamino) do not correlate well at all (*21*) but appear to be significantly less electron withdrawing than their σ^* values would suggest. This has been attributed to $(p \rightarrow d)\Pi$ back donation from the substituent into the vacant 3*d* orbitals of silicon (*21, 22*), resulting in partial cancellation of the inductive electron-withdrawing power of the group.

In this research, since the silanes included mono-, di-, and trihydro species, the sum of *all four* σ^* values was taken as indicative of the total electronic environment of silicon. When the logarithms of the relative rates ($\log k_{rel}$) were plotted against $\overset{4}{\Sigma}\sigma^*$ (Table I and Figure 1), an excellent correlation was observed for most substituents (*vide infra*). Least-squares analysis of the data afforded Equation 1.

$$\log k_{rel} = -1.2513 \overset{4}{\Sigma} \sigma^* + 2.2166 \qquad (1)$$

Substituents whose standard Taft σ^* values produced points far removed from the plot of Figure 1 included phenyl, chloromethyl, di-

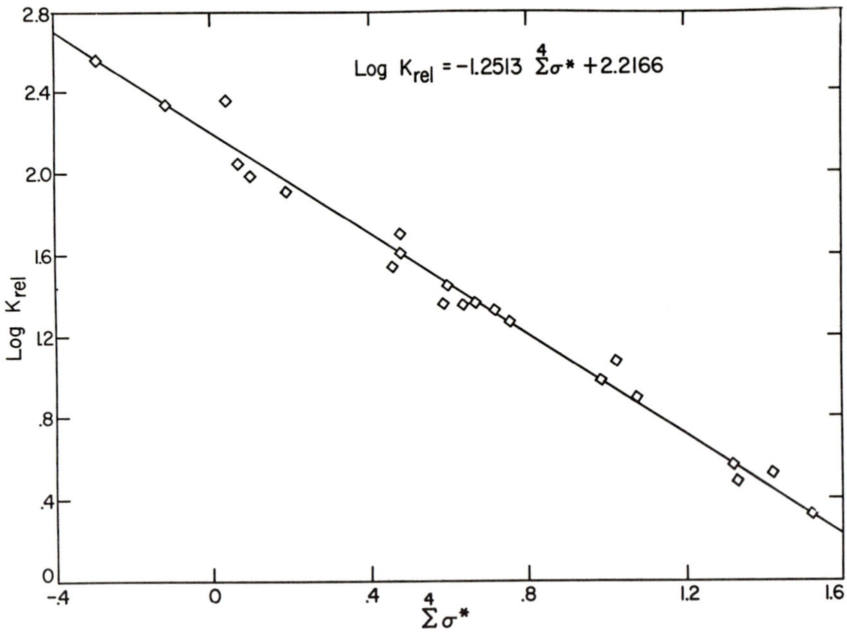

Figure 1. Hydrosilane–ozone rate/$\overset{4}{\Sigma}\sigma^*$ *correlation*

Table II. New σ_{Si}^* Values

R	σ_{Si}^{*a}	σ^{*f}	$\sigma^* - \sigma_{Si}^*$
1-$C_{10}H_{17}$– [b]	−.26 [c]		
$F_3C–CH_2–CH_2$–	−.01	0.32	0.33
Cl_2CH–	−.01	1.94	1.95
2,2′-Biphenylenyl	.01		
Phenyl	.05 [d]	0.60	0.55
C_2H_5O–	.09 [e]	1.35	1.26
$ClCH_2$–	.18	1.05	.87

[a] This paper.
[b] Mixture of 1-decalinyl isomers obtained by catalytic hydrogenation of 1-naphthyl on silcon (*5*).
[c] This value is probably well behaved and equal to σ^*.
[d] In exact agreement with value of Attridge.
[e] Substantially different from calculated value of 0.75 for CH_3O–, whose standard σ^* is 1.35 (*see* text).
[f] Standard Taft values (20, 29).

chloromethyl, ethoxy, and 3,3,-trifluoropropyl. These points may be brought into correlation by redefining the substituent constant as σ_{Si}^*, whose value may be deduced from k_{rel}. Table II lists the values of σ_{Si}^* thus derived. The factor which these substituents have in common is the presence of unshared electron pairs or Π-electrons close to the silicon atom.

The value of σ_{Si}^* for phenyl, 0.05, agrees with that calculated for phenyl from the Si–H stretching frequency (*21*). This discrepany between the latter and 0.60, the "normal" σ^* for phenyl, has also been attributed to $(p \rightarrow d)\Pi$ back bonding (*21, 23–25*). However, the here-observed σ_{Si}^* for ethoxy, −0.09, does not agree with the value of 0.75 derived from ν_{Si-H} stretch for methoxy although their σ^* values are quite similar—1.35 and 1.45, respectively. Indeed, all of the "electron-rich" substituents indicate a near cancellation of inductive and mesomeric effects. However, alternate explanations of the so-called $(p \rightarrow d)\Pi$ back donation exist (*26, 27*).

Improved understanding and treatment of the phenyl case was developed from the recognition that the derived σ_{Si}^* (C_6H_5–) is almost identical with the "standard" Hammett σ value, a parameter containing both mesomeric and inductive contributions. Since published data on the k_{rel} of ozonation of substituted phenyldimethylsilanes were available (*11*), and included the common point, phenyldimethylsilane, simple mathematical manipulation could bring these and the present data on to the same scale. When the Hammett σ for the substituted phenyls was used in the $\overset{4}{\Sigma}\sigma^*$ calculation, all of the compounds correlated well with Equation 1. A comparison of the found σ_{Si}^* (X–C_6H_4–) and their Hammett σ values (*29*) appears in Table III.

In essence, then, the electronic environment of the silicon can be described by a substituent value, σ_{Si}^*, which is (a) equal to the Taft σ^* if the substituent is an alkyl group, (b) equal to the Hammett σ if the substituent is phenyl or substituted phenyl, and (c) for substituents bearing unshared electron pairs close to the silicon, each σ_{Si}^* must be individually determined. The σ_{Si}^* values computed in this study appear in Table II.

Reaction Mechanism. The proposed mechanism for the ozone–hydrosilane reaction (7) shown in Equation 2, as deduced by analyzing and correlating data on relative rates, substituent effects, deuterium isotope effects, low temperature NMR, and ultraviolet spectroscopy for a range of hydrosilanes, is a multistep one as follows:

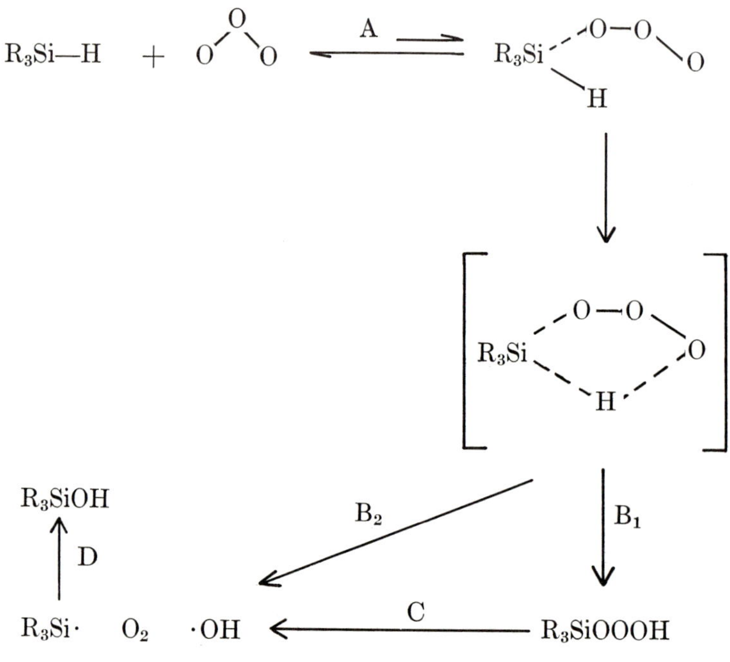

(2)

In the next sections of this report, each step of this reaction scheme is considered in turn, the alternatives discussed, and the rationale for decision presented.

Step A—Association of Ozone with the Silicon Atoms. The linear Hammett-type relationship of Figure 1 and Equation 1 indicates a slope, ρ, of -1.25. This negative value denotes electrophilic attack by the ozone and/or the development of a partial positive charge on silicon in the transition state. Since the silicon is relatively electropositive, an electrophilic attack by ozone on silicon seems unlikely. The hydrogen bound to silicon, however, is hydridic in character and is the likely site of attack by an electrophile. If attack by ozone on hydrogen occurs in the rate step, a primary deuterium isotope effect is predicted (*30*).

The deuterium isotope effect, k_H/k_D, was determined for the reaction of tri(*n*-butyl)silane with ozone, by comparing the k_{rel} for the oxidation of $(n\text{-}C_4H_9)_3SiH$ and $(n\text{-}C_4H_9)_3SiD$ *vs.* $(n\text{-}C_6H_{13})_3SiH$. Ozonation here exhibits a surprisingly large primary isotope effect, the ratio of the proto- to the deuteriosilane, k_H/k_D, being 6.9. The calculated k_H/k_D, based on the zero-point energy effect using the infrared stretching frequencies of the Si–H and Si–D bonds in tri(*n*-butyl)silane, is 4.5 at 0° (*31*). This large isotope effect is indicative of the involvement of Si–H bond breaking in the rate step and, therefore, corroborates electrophilic attack by ozone on the hydridic hydrogen. The possibility that this large isotope effect could be caused by autoxidation by the oxygen carrier gas, as observed in aldehydes (*12*), has been ruled out since the k_H/k_D is unchanged if argon is used as the ozone carrier rather than oxygen (*10*).

In view of the possibility that attack by ozone may proceed *via* oxygen insertion into the Si–H σ-bond (state **1**), the known reaction of dichlorocarbene with the Si–H bond (*32*) was taken as a model for such insertion. This reaction is postulated to proceed *via* the three-center transition state **2**, which is structurally analogous to **1**. The isotope effect, k_H/k_D, for insertion of dichlorocarbene into $(n\text{-}C_4H_9)_3SiH$ and $(n\text{-}C_4H_9)SiD$, determined in the same manner as for their ozonations, was found to be 1.23 (*10*). The large difference in the k_H/k_D values

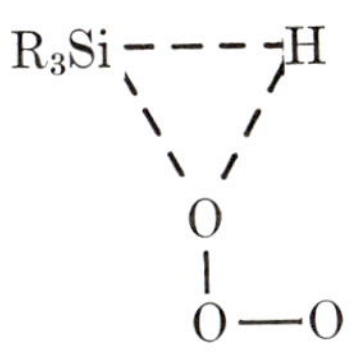

1

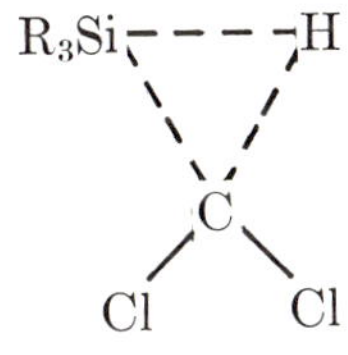

2

suggests that the mechanism of the ozone reaction does not closely resemble that for the carbene insertion. Furthermore, a Hammett-type $\rho\sigma$ treatment of the carbene insertion produced a ρ value of -0.63 (*32*) compared with -1.25 for the ozone reaction. This indicates a greater separation of charge (less concertedness) in the latter although both do involve electrophilic attack.

The primary isotope effect and the relative rate data, however, do conflict at one point. If attack on hydrogen is, indeed, rate determining, one would expect a statistical factor for attack on the di- and trihydrosilanes. These compounds are, however, correctly accommodated on the same linear plot (Figure 1) as are the monohydrosilanes.

Table III. σ_{Si}^* Values for Substituted Phenyl Groups

X in XC_6H_4–	(k_x/k_H) [a]	$\sum^{4}\sigma_{Si}^*$ [b]	$\sigma_{Si}^*(XC_6H_4\text{–})$ [c]	σ [d]
p–CH_3O	2.5	.22	−.27	−.268
p–CH_3	1.8	.34	−.15	−.170
H	1.0	.54	.05	0.000
p–Cl	0.48	.80	.31	.227
m–Cl	0.31	.87	.38	.373
p–CF_3	0.20	1.03	.54	.54

[a] Relative rate of ozonization of substituted phenyldimethylsilane compared with phenyldimethylsilane (*11*).

[b] Calculated from ozonization rates of aryldimethylsilanes (*11*) relative to phenyldimethylsilane and use of equation of Figure 1, assuming .05 for σ_{Si}^* (phenyl).

[c] Obtained by subtraction, from preceding column, of σ^* values for hydrogen and two methyls.

[d] Standard Hammett substituent constants based on ionization constants of substituted benzoic acids (28).

One possible explanation for this behavior is that charge separation occurs in the transition state (producing either $\equiv Si^+$ or $\equiv Si\cdot$) and that the species order of stability (tertiary > secondary > primary) exactly cancels the statistical factor. Although this explanation should not be rejected out-of-hand, the fortuitous coincidence required makes it unattractive.

A second possibility is that the ozone forms some kind of complex with the silane before attack on the hydrogen. From this complex, all hydrogens are equally accessible, and the decomposition is first order in complex. In the hope of observing such a complexation, the ultraviolet spectra of ozone/silane mixtures in carbon tetrachloride were examined (*33*). Although no spectral bands attributable to a silicon–ozone complex were found, it was observed that *any* silicon-containing species catalyzed the decomposition of ozone. That is, not only triethylsilane, but triethylsilanol and tetramethylsilane as well, destroy ozone in carbon tetrachloride. This result indicates an association of the ozone with the silicon atom, regardless of the functionality of the silicon species (within the types examined) and completely independent of the silicon substrate's

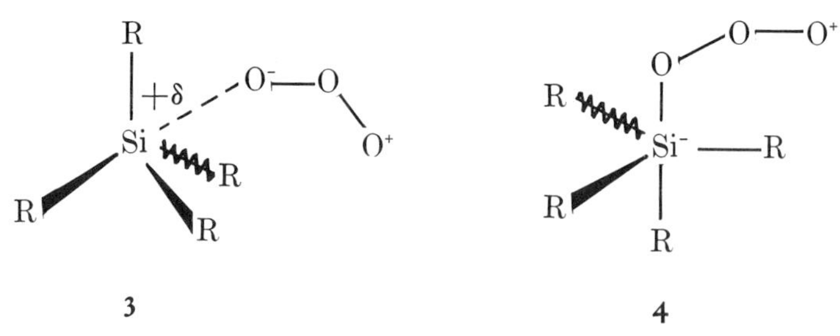

3 4

ability to enter into a subsequent transformation reaction. In view of the electropositive nature of silicon, the nucleophilic association of ozone with the silicon is the likely candidate for a suitable intermediate. Possible models for this type of interaction range from one of weak electrostatic attraction, **3**, to a fully pentacoordinate silicon species, **4**.

In an attempt to observe the complex, a low temperature NMR investigation of mixtures of ozone with triethylsilanol and with tetramethylsilane were conducted. At −57°C in methylene chloride, there was no discernible change in the proton NMR of either triethylsilanol or tetramethylsilane as the ozone concentration of the resulting blue solutions was increased to the point of saturation. Two possible explanations for these results are: (a) the complex exists in equibrium with the silane and ozone, and the actual concentration of complex is too small to be detected by NMR (less than 5% in this experiment), or (b) the chemical shifts of the silane protons are not altered by complexation. Of the two explanations, the first seems the more reasonable, particularly if the complex has more of the character of **4** than of **3**. The exact nature of the complex is impossible to determine at this time, but one significant observation is that optically active perhydro-1-naphthylphenylmethylsilanol undergoes slow racemization upon standing at room temperature in a pentane–ozone solution. This is consistent with reversible complexation to a species having some pentacoordinate character from which racemization may occur *via* pseudorotation (*34*).

These data suggest that the initial step in the reaction of ozone with the Si–H bond is the reversible formation of a silicon–ozone complex. This cannot be the rate step since ρ would have to be positive (nucleophilic attack) and no primary isotope effect would be predicted. To eliminate the statistical factor for the di- and trihydrosilanes, attack by ozone on the hydridic proton from within the complex must be much more favorable than direct encounter and reaction with uncomplexed ozone and Si–H.

Step(s) B_1 and B_2—Hydrogen Abstraction. How the ozonation proceeds from the complex is now considered. If direct σ-bond insertion (*1*) is eliminated on the basis of isotope effects as discussed above, then **5** and **6** are the viable alternatives. The transition state **5** could collapse to form the silicon hydroperoxide **7**, while transition state **6** could collapse to form the silicon hydrotrioxide **8** (path B_1); alternatively, **6** could collapse directly to ion or radical pairs (path B_2). The transformation **5** → **7** is not meant to suggest that atomic oxygen is the other product. Since the reaction order in ozone has not been determined, the fate of the other oxygen atom(s) is moot.

5 7

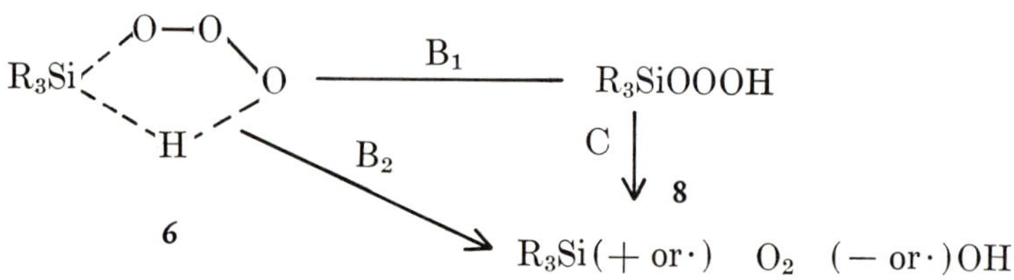

6 8

If the reaction is four-centered (**5** → **7**), a silicon hydroperoxide should be formed. The reaction of triethylsilane with ozone was monitored by NMR at −57°C. The only species observed under these conditions were the silane and the silanol; no evidence for a hydroperoxide intermediate which might have been stable at that temperature (*35*) was detected, and chemical tests for peroxides proved negative (*5*, *6*). As the concentration of ozone was increased from zero to saturation, the spectrum of the silane completely disappeared with the concurrent appearance of the silanol NMR spectrum.

If the reaction is five-centered (**6** → **8**), a silicon hydrotrioxide may be formed. Since there has been no previous report of a silicon hydrotrioxide, the stability of a species such as **8** under these conditions can only be estimated. Among carbon analogs, the reported dialkyltrioxides have only marginal stability at low temperatures (*36*, *37*), and alkyl hydrotrioxides, proposed as intermediates in the ozonation of alcohols and ethers (*38*), decompose at *ca.* −10°C. Admittedly, speculative extrapolation based on the comparative stabilities of other types of silicon and carbon analogs suggests that a silicon hydrotrioxide **8**, should have been observable if it had been present, but this requirement is debatable.

Steps C and D—The Nature of the Recombining Fragments. Upon assumption of the intermediacy of a five-center transition state, the question then arises as to how **6** (or **8**) decomposes to the silanol. The breakdown could be *via* radical or ion pairs. The large primary isotope effect suggests a radical pathway (*12*). Further insight was gained from the ozonation of optically active perhydro-1-naphthylphenylmethylsilane, which yields the silanol **10** with retention of configuration (Equation 3). Prolonged exposure of **10** to ozone causes racemization, but the product

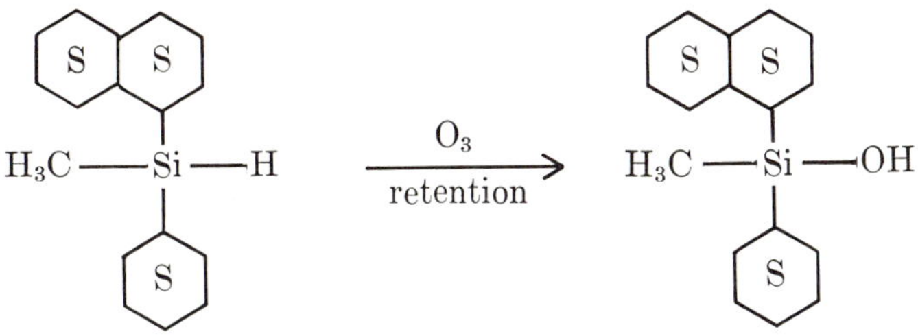

(3)

9 10

initially has retained its configuration. Presumably, the silane **9** is also racemized by ozone (*via* complexation), but the rate of ozonization is faster. Retention of configuration by silyl radicals has been demontrated (*39*) whereas the optical properties (indeed, the very existence) of trivalent siliconium ions have not been established. For this reason, decomposition of **6** (or **8**) through a radical pair with subsequent, rapid recombination is the mechanism which appears most compatible with the experimental observations.

Conclusions

When all the mechanistic evidence is taken into consideration, the following reaction sequence appears to best satisfy the data. The silane undergoes reversible complexation (A) with the ozone, the complex being present in only small concentrations. The rate step then involves electrophilic attack on the hydridic hydrogen, passing through a five-center transition state. This may decompose to either a silyl hydrotrioxide (B_1) or directly to the radical pair (B_2). The silyl hydrotrioxide, if present, must decompose rapidly to the radical pair (C). This radical pair then recombines with retention of configuration to afford the ultimate product, the silanol (D).

With regard to the five-center transition state, Kenneth Wiberg (Yale University) has kindly pointed out to us that the magnitude of the isotope effect supports the five-centered activated complex which we have postulated. In the smaller ring complexes, the vibrational mode converted to the reaction coordinate is a bending mode, which would not produce a large isotope effect. In the five-centered complex, both bending and stretching modes are converted, and the large isotope effect is not unexpected.

Finally, a referee has suggested that **8** could decay to product *via* a concerted pathway (i) [p. 76]. Although such a pathway cannot be ruled out, we feel that the large isotope effect suggests the radical pair, probably from B_2. Both i and the radical mechanism would proceed with retention of configuration.

$$\begin{array}{l} R_3Si{-}O{-}O \rightarrow R_3SiOH + O_2 \\ \qquad\quad\;\; | \\ \qquad\;\; H{-}O \end{array} \qquad \text{(i)}$$

Literature Cited

1. Long, L. Jr., *Chem. Rev.* (1940) **27,** 437.
2. Bailey, P. S., *Chem. Rev.* (1958) **58,** 925.
3. Barry, A. J., Beck, H. N., "Inorganic Polymers," pp. 221, 236, 256, 296, Academic, New York, 1962.
4. Spialter, L., Austin, J. D., *J. Amer. Chem. Soc.* (1965) **87,** 4406.
5. Austin, J. D., Spialter, L., ADVAN. CHEM. SER. (1968) **77,** 26–31.
6. Spialter, L., Austin, J. D., *Inorg. Chem.* (1966) **5,** 1975.
7. Spialter, L., Pazdernik, L., Bernstein, S., Swansiger, W. A., Buell, G. R., Freeburger, M. E., *J. Amer. Chem. Soc.* (1971) **93,** 5682.
8. Spialter, L., Swansiger, W. A., *J. Amer. Chem. Soc.* (1968) **.90,** 2187.
9. Spialter, L., Swansiger, W. A., "Resumes of Communications," Second International Symposium on Organosilicon Chemistry, Bordeaux, pp. 183–184, 1968.
10. Spialter, L., Swansiger, W. A., Pazdernik, L., Freeburger, M. E., *J. Organometal. Chem.* (1971) **27,** C25.
11. Ouellette, R. J., Marks, D. L., *J. Organometal. Chem.* (1968) **11,** 407.
12. Batterbee, J. E., Bailey, P. S., *J. Org. Chem.* (1967) **32,** 3899.
13. Erickson, R. E., Hanson, R. T., Harkins, J., *J. Amer. Chem. Soc.* (1968) **90,** 6777.
14. Erickson, R. E., Bakalik, D., Richards, C., Scanlon, M., Huddleston, G., *J. Org. Chem.* (1966) **31,** 461.
15. White, H. M., Bailey, P. S., *J. Amer. Chem. Soc.* (1965) **30,** 3037.
16. Becsey, J., Berke, L., Callan, J. R., *J. Chem. Ed.* (1968) **45,** 728.
17. Ponomarenko, V. A., Egorov, Yu P., *Izvest. Akad. Nauk S.S.S.R., Otdel. Khim. Nauk* (1960) 1133.
18. Smith, A. L., Angelotti, N. C., *Spectrochim. Acta* (1959) **15,** 412.
19. Thompson, H. W., *Spectrochim. Acta* (1960) **16,** 238.
20. Newman, M. S., Ed., "Steric Effects in Organic Chemistry," p. 619, Wiley, New York, 1956.
21. Attridge, C. J., *J. Organometal. Chem.* (1958) **13,** 259.
22. Egorochkin, A. N., Vyazankin, N. S., *Dokl. Akad. Nauk S.S.S.R.,* (1970) **193,** 590.
23. Alt, H., Bock, H., *Tetrahedron* (1969) **25,** 4825.
24. Bock, H., Alt, H., *Chem. Ber.* (1970) **103,** 1784.
25. Bock, H., Alt, H., *J. Amer. Chem. Soc.* (1970) **92,** 1569.
26. Pitt, C. G., *J. Organometal. Chem.* (1970) **24,** C35.
27. Ramsey, B. G., "Electronic Transitions in Organometalloids," pp. 76ff, Academic, New York, 1969.
28. McDaniel, D. H., Brown, H. C., *J. Org. Chem.* (1958) **23,** 420; *see also* Ref. *29*, p. 87.
29. Hine, J., "Physical Organic Chemistry," 2nd ed., p. 97, McGraw-Hill, New York, 1962.
30. Westheimer, F. H., *Chem. Rev.* (1961) **61,** 265.
31. Melander, L., "Isotope Effects on Reaction Rates," p. 20, Ronald Press, New York, 1960.
32. Seyferth, D., Damrauer, R., Mui, J. Y.-P., Jula, T. F., *J. Amer. Chem. Soc.* (1968) **90,** 2944.

33. Nakagawa, T. W., Andrews, L. J., Keefer, R. M., *J. Amer. Chem. Soc.* (1960) **82**, 269.
34. Muetterties, E. L., *Accounts Chem. Res.* (1970) **3**, 266.
35. Dannley, R. L., Jalics, G., *J. Org. Chem.* (1965) **30**, 2417.
36. Bartlett, P. D., Guaraldi, G., *J. Amer. Chem. Soc.* (1967) **89**, 4799.
37. Mill, T., Stringham, R. S., *J. Amer. Chem. Soc.* (1968) **90**, 1064.
38. Murray, R. W., Lumma, W. C., Lin, J. W.-P., *J. Amer. Chem. Soc.* (1970) **92**, 3205.
39. Brook, A. G., Duff, J. M., *J. Amer. Chem. Soc.* (1969) **91**, 2118.

RECEIVED May 20, 1971.

7

The Ozonolysis of Organomercurials

WILLIAM L. WATERS, PAUL E. PIKE, and JOSEPH G. RIVERA

University of Montana, Missoula, Mont. 59801

The reaction of ozone with over 20 different dialkylmercurials and alkylmercuric halides produced good yields of carboxylic acids, ketones, and alcohols from primary, secondary, and tertiary alkylmercurials, respectively. Although carbon–mercury scission was the chief course of reaction, some carbon–carbon cleavage did occur in varying degrees. A kinetic study of the reaction under pseudo first-order rate conditions indicated a build-up of positive charge on carbon in the rate-determining step. The mechanism for carbon–mercury cleavage is depicted as a mixture of pure S_E2 and four-center processes. Although somewhat limited in scope, ozonolysis of organomercurials does promise synthetic utility.

In the search for new synthetic uses of organomercurials and the olefin mercuration reactions in general, attention was drawn to ozone as a possible reactant.

Organomercurials react rapidly with a variety of electrophilic species, and a recent book by Jensen and Rickborn fully describes the scope of these reactions (*1*). Foremost among the electrophilic reagents are the halogens which cleave the carbon–mercury bond to produce alkylmercuric halides and/or alkyl halides, depending on the type of organomercurial involved.

$$\begin{array}{l} R_2Hg + Br_2 \xrightarrow{\text{fast}} RBr + RHgBr \\ RHgBr + Br_2 \xrightarrow{\text{slower}} RBr + HgBr_2 \end{array} \qquad (1)$$

In addition, protic acids also cleave organomercurials to yield the proton-substituted products analogous to those for halogenation above. Finally, mercuric salts also act as electrophilic reagents towards organomercurials.

$$R_2Hg + HgBr_2 \longrightarrow 2RHgBr$$

$$RHgBr + {}^{*}HgBr_2 \longrightarrow R^{*}HgBr + HgBr_2 \qquad (2)$$

$$\text{where } {}^{*}Hg = {}^{203}Hg$$

Several mechanisms have been proposed for these electrophilic cleavage reactions. In general, these mechanisms range from pure S_E1 to pure S_E2 processes although these two extreme mechanisms are usually reserved for special cases.

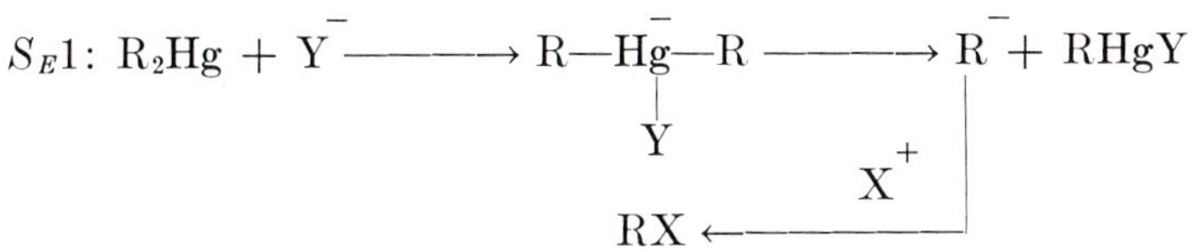

(3)

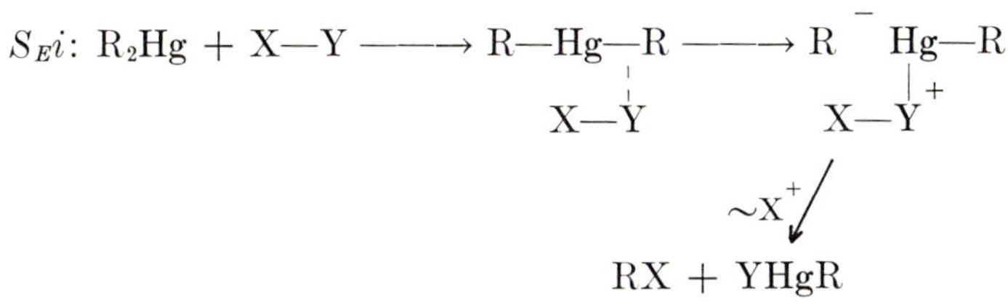

(4)

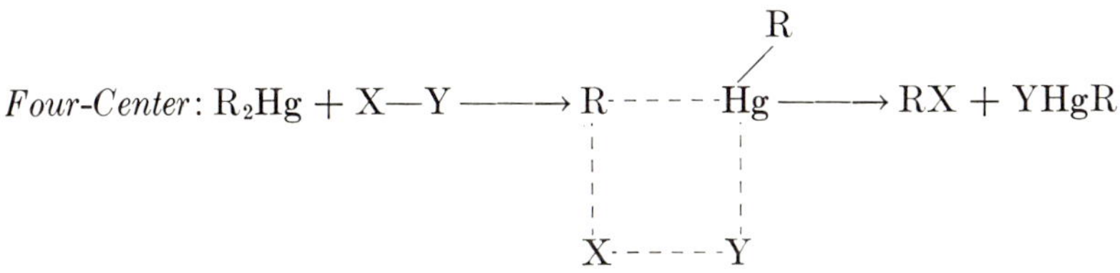

(5)

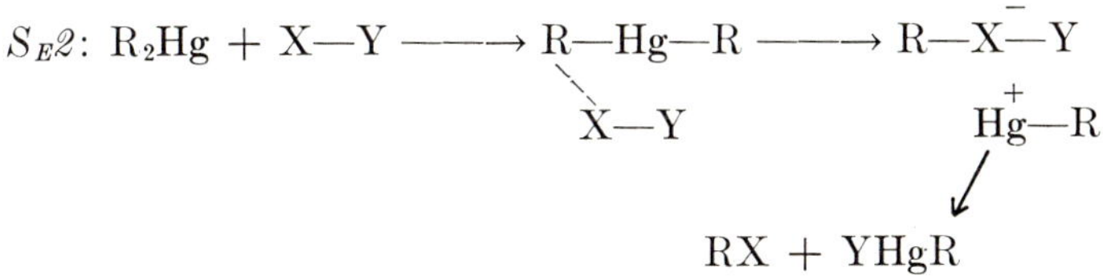

(6)

Ozone has long been assigned the role of electrophile in addition and substitution reactions (*2*). In this respect the reagent has been reported to clcave various carbon-metal bonds (*3–9*).

$$(Et)_4Sn \xrightarrow[\text{nonane}]{O_3} AcH + (Et)_2SnO + \text{a peroxide} \qquad (7)$$

$$(Et)_4Pb \xrightarrow[20^\circ]{O_3} (Et)_3PbOH + (Et)_2PbO + (Et)_3PbOEt + AcH + EtOH \quad (8)$$

$$(Et)_4Si \xrightarrow{O_3} AcOH + H_2O_2 + Et_3SiO_2H + (Et_3Si)_2O \quad (9)$$

$$R_2Se \xrightarrow{O_3} R_2SeO \longrightarrow RSeH + \text{aldehyde} \quad (10)$$

$$(i\text{-}Pr)_2Hg \xrightarrow{O_3} 2(CH_3)_2CO + HgO \quad (11)$$

Although reports of mechanistic studies are scarce for these reactions, Ouellette and Marks did propose a 1,3-dipolar attack by ozone (*10*). The corresponding intermediate was predicted to be a hydrotrioxide of silicon.

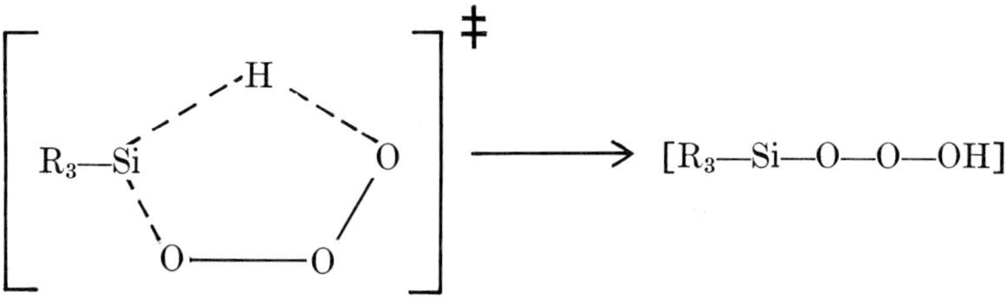

(12)

The unstable hydrotrioxide could then decompose *via* singlet or triplet oxygen ejection to form the observed silanol products (*10*).

The Reaction

The cleavage of carbon–metal bonds by ozone is intriguing. Study was begun to discover the scope of this reaction using mercurials as the organometallic reagent. It was hoped that not only would the cleavage mechanism be clarified but also that a synthetically useful method of introducing oxygen functionality might be developed.

Ozonation of over 20 different organomercurials resulted in a fairly rapid disappearance of organometallic starting material, accompanied by a constant precipitation of inorganic mercury salts. A product study in chloroform and methylene chloride solvents at temperatures ranging from −75° to +25°C showed that mercury was replaced by an oxygen-containing moiety in all cases (*11*). Although these products were chiefly the analogous alcohol, ketone, or acid from the parent 3°, 2°, or 1° organomercurial respectively, some carbon–carbon scission did occur.

This was especially evident at the higher temperatures and in the case of the alkylmercuric halides.

A kinetic study of the reaction was also performed in which NMR-obtained rate data were correlated with mercurial structure changes (*12*). This study revealed a quite distinct reactivity order which, coupled with a 1:1 reactant stoichiometry, indicates a 1,3-dipolar electrophilic attack by ozone *via* a S_E2 or four-center process. Although the exact mechanism was not conclusively proved, it is certain that neither the S_E1 or S_Ei processes were operative during these reactions.

Experimental

Materials. Generally the organomercuric halides were prepared using the Grignard method as outlined by Marvel *et al.* (*13*). The dialkylmercurials were also synthesized by the Grignard procedure as outlined by Gilman and Brown (*14*). Exact structural identification of the organomercurials was facilitated by correlation of NMR-obtained $^{199}Hg-^{1}H$ spin-spin coupling data with literature values for similar systems (*15, 16*). Preliminary identification was performed by the usual comparison of physical properties with those for known compounds.

Apparatus. The apparatus used in these studies consisted of a Welsbach T-408 electric discharge ozonator connected to three 200-ml gas washing bottles in series. The first bottle was the reaction vessel, and its temperature was maintained at $-76°$, $0°$, or $+10°C$. The second bottle was a cold trap for the more volatile products—*e.g.*, acetone—which were generally swept out of the reaction vessel by the large volume of O_2. The third bottle contained an aqueous KI solution for trapping any unreacted O_3. After completion of a reaction this solution could be titrated with thiosulfate for liberated I_2 and thus unreacted O_3. By using O_2 as the feed gas, an O_3–O_2 mixture, 3-4% O_3 by weight, was produced by the ozonator. Flow rates were adjusted to produce 18–30 mmoles of O_3 per hour. For the rate studies the above-mentioned reaction vessel was modified to allow for rapid and periodic removal of solution aliquots during the progress of the reaction.

Procedure. PRODUCT-DETERMINING STUDIES. In a typical product-determining run 10 mmoles of organomercurial were dissolved in *ca.* 50 ml of CH_2Cl_2 and placed in the reaction vessel. After the vessel was cooled to the proper temperature, the O_3–O_2 flow was started. In all cases an immediate white precipitate appeared. This precipitation was constant throughout the entire ozonolysis procedure and in fact was used as an endpoint determinant in certain cases.

Ozonation was generally halted after all of the organomercurial had reacted. The resultant white solid was then collected, washed with CH_2Cl_2, dried, and subjected to analysis by powder x-ray diffraction with a Norelco analytical x-ray diffractometer.

The CH_2Cl_2 solution was analyzed by vapor phase chromatography (VPC) methods using 6 ft × ¼ inch 30% Carbowax or 6 ft × ¼ inch 20% SE-30 columns. Retention times of the reaction products were matched on both columns with precalibrated chromatographs of known compounds. An internal standard was added to the CH_2Cl_2 solution for integration purposes.

KINETIC STUDIES. In a typical kinetic run 10 mmoles of organomercurial were dissolved in 100 ml of $CHCl_3$ and placed in the reaction vessel. To this solution were added 30 mmoles of CH_3NO_2 as an internal NMR standard. Ozone flow rates were determined and the mercurial solution was then ozonized at 0°C. Aliquots were periodically withdrawn (generally, 8–10 samples taken) and shaken with a solid mixture of *ca.* 0.1 gram KI and 0.1 gram $Na_2S_2O_3$. In this manner excess O_3 was immediately quenched and the resultant I_2, which itself reacts with organomercurials at room temperature, was quickly reduced.

Each aliquot was analyzed without further work-up by an HA-60 Analytical NMR Spectrometer, using the field sweep method to eliminate the usual phase changes associated with the frequency sweep mode. The relative integrals for CH_3NO_2 and one set of resonance lines for the organomercurial were repeated 10 times and averaged. This average value was compared with the original CH_3NO_2/organomercurial ratio to determine the concentration of organomercurial at that time. These actual concentrations, along with the respective time intervals from t_0, were submitted to a least-squares computer program. Using an IBM 1620 computer, the natural logarithm of the concentration was plotted *vs.* time (*ca.* 10 points) to give the best-fitting linear relationship between the two variables. From this plot the least-squares slope, *y*-intercept, absolute rate constant, and correlation coefficient of the line were determined.

OTHER STUDIES. Certain ozonations were carried out using N_2 as the carrier gas instead of O_2. For these reactions the usual silica gel procedure (*17*) was used. Silica gel, 100–150 grams, 6–12 mesh, was placed in a 200-ml side-arm cold trap and cooled to −76°C. Ozone was introduced into this trap as the usual O_3–O_2 mixture. After the desired amount of O_3 had been adsorbed by the silica gel, the ozonator was disconnected and the cold trap flushed with ambient temperature N_2 for 5 minutes to remove any O_2. Ozone was then flushed slowly from the silica gel by gradually moving the trap very short distances out of the dry ice—acetone bath while continuing to flush with N_2 [hastening this warming procedure resulted in a violent explosion].

Next, a stoichiometric study was undertaken to determine the number of moles of O_3 which reacted with each mole of organomercurial. To accomplish this task the O_3 flow rate was accurately determined by the KI–$Na_2S_2O_3$ procedure. Ozonation of a mercurial in CH_2Cl_2 or $CHCl_3$ was then begun. Included in the reaction mixture was a small amount of CH_3NO_2 as an NMR integration standard. The reaction was stopped several times to check the organomercurial concentration by NMR spectroscopy. The amount of unreacted ozone also had to be calculated at each sampling period for the primary and secondary alkylmercuric halides. The O_3 flow rate was also rechecked at each sampling interval to ensure that the dispersion tubes had not clogged. The number of moles of reacted O_3 was then compared with the number of moles of reacted organomercurial.

Results

Product Studies. Table I outlines the organic and inorganic products of the ozonolysis reactions. The ozonolysis products of *n*-alkylmercuric halides and di-*n*-alkylmercurials were a mixture of carboxylic acids which had a carbon chain equal to or shorter than the alkyl group of the parent organomercurial.

Ozonolysis of the *s*-alkylmercuric halides and the di-*s*-alkylmercurials produced the coresponding ketone. Although some carbon–carbon cleavage occurred, it was generally less than with the reaction of the primary organomercurials (*see* Reactions *13* and *14,* Table I). In partial contrast to the results of Bockemuller and Pfeuffer (Reaction *11*), the ozonation of diisopropylmercury yielded acetic acid in addition to acetone (Reaction *14,* Table I).

Tertiary alkylmercuric halides yielded a considerable amount of the corresponding alcohol upon ozonation (Reaction *17,* Table I). Some carbon–carbon cleavage accompanied the main reaction in this case as well.

The inorganic products of the ozonolysis reactions were determined for three different organomercurials. Ozonolysis of two dialykylmercurials produced a mixture of mercuric chloride, mercurous chloride, and mercuric oxide (Reactions *3* and *14,* Table I) while one alkylmercuric halide gave only mercuric and mercurous chlorides (Reaction *13,* Table I). A known mixture of the three salts was tested for its stability to the reaction conditions. The salts were ozonized as a solution/mixture with methylene chloride. Powder x-ray diffraction showed no difference in the mercury salt mixture after a 2-hour ozonation at 10°C.

Carbon–Carbon Cleavage. As mentioned above, carbon–carbon cleavage usually accompanied the "normal" carbon–mercury cleavage reaction. While this chain-shortening process was most prominent for the *n*-alkylmercuric halides which were ozonated at 10°C (Reactions *1* and *6,* Table I), it was reduced to a much lower level when the di-*n*-alkylmercurials were ozonated at −76°C (Reactions *5* and *10,* Table I). Actually even less carbon–carbon scission occurred during the ozonation of the di-*s*-alkylmercurials halides (Reaction *14,* Table I) and during the partial ozonation of the *tert*-alkylmercuric halides (not listed).

By comparison of the *n*-alkylmercurial results alone, it is possible to generalize that both higher ozonation temperatures and the presence of a halogen ligand on mercury promote carbon–carbon cleavage during the ozonolysis of the carbon–mercury bond.

As a check of a possible reaction pathway for the formation of the shorter acids during ozonation of *n*-alkylmercurials, pentanoic acid was ozonized for 4 hours at 10°C. The ozonation was carried out in dichloromethane which contained small amounts of mercuric chloride and mercurous chloride. Only starting material could be detected by VPC analysis of the product mixture.

In another test of the reaction pathway for carbon–carbon cleavage, the exit gas from the reaction vessel during a 10°C ozonation of *n*-propylmercuric bromide was passed through a $Ba(OH)_2$ solution. This solution showed an almost immediate and constant precipitation of $BaCO_3$ during ozonation. However, further investigation revealed that methylene chloride yielded CO_2 at about the same rate at 10°C. Hence, after determining that Freon 114 did not react with ozone at −76°C, the test was repeated on di-*n*-propylmercury at the reduced temperature and in the inert solvent. Under these conditions CO_2 was still produced at an appreciable rate. One final experiment, however, placed the origin of the CO_2 back into an undecided light. Formic acid was oxidized slowly to CO_2 when ozonized in Freon 114 at −76°C.

The Role of Oxygen. To define properly the role which O_2 played in these ozonolysis reactions, nitrogen was occasionally used as the carrier gas for ozone. After ozonizing *n*-hexylmercuric bromide at 10°C with the usual O_3–O_2 mixture, the reaction was repeated with O_3–N_2 (*see* Experimental). By comparing the reaction products from runs *6* and *7* (Table I), it can be readily seen that no product differences exist when N_2 is substituted for O_2.

In a significantly different approach to the problem, isopropylmercuric chloride was oxygenated in $CHCl_3$ at 10°C for 2 hours at 24 liters of O_2 per hour. Analysis (VPC) showed that 4% of the organomercurial had reacted to give acetone.

Stoichiometry of Ozone Uptake.. The results of the stoichiometry study are outlined in Table I.

Intermediate Products from Ozonolysis of Organomercurials. A bonus of the NMR method for kinetic analysis of the O_3–organomercurial reaction was the fact that certain O_3-reactive intermediates were observed during ozonolysis of some of the organomercurials.

Diisopropylmercury yielded isopropylmercuric chloride and isopropyl alcohol as well as acetone. The first two products gave acetone upon further reaction with O_3. The following equation appears to describe this system:

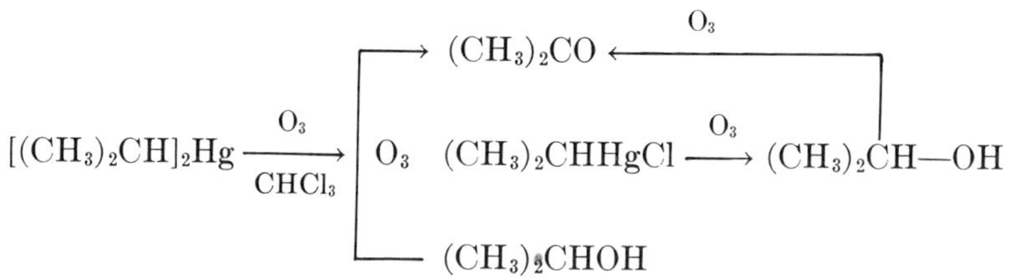

(13)

The belief that acetone was one of the initial products of the reaction is supported by its extremely fast rate of formation compared with the rate of oxidation of isopropyl alcohol to acetone by O_3–O_2.

A study of the intermediate products from the ozonolysis of isoproplymercuric chloride shows that the rate of build-up of isopropyl alcohol adequately accounts for the rate of acetone formation. The reaction sequence therefore appears to be one of carbon–mercury cleavage, followed by oxidation by ozone.

The tertiary alkylmercuric halide, *tert*-butylmercuric chloride yielded only *tert*-butyl alcohol as the primary product. It was not until all the organomercurial had reacted that acetone appeared as a reaction product (*see* Reaction *17*, Table I). Thus it again is suggested that an alcohol is the immediate product of C–Hg cleavage by ozone, at least in the case of alkylmercuric halides.

Results

Kinetic Studies. The ozonation rates of 10 alkylmercuric halides were determined in chloroform at 0°C. The rate of ozone oxidation of isopropyl alcohol to acetone under identical conditions was also measured. An attempt to measure the ozonation rate of six dialkylmercurials failed, however, because the reaction with R_2Hg was much too fast under these conditions.

The previously described method used to obtain rate constants for these reactions involved the continuous passage of a stream of O_3–O_2 through the reaction vessel. For the kinetic study of the slower-reacting alkylmercuric halides the O_3 concentration remained saturated and therefore constant. This was not true for the dialkylmercurials, however, since they reacted so fast with ozone that an O_3 concentration approaching the saturation value could never be reached. This situation was also borne out by the consistent observation that the KI trap (of the reaction apparatus) became colored during the dialkylmercury ozonations only after all of the mercurial had reacted. By comparison, the reactions of the alkylmercuric halides were slow enough so that iodine was formed in the KI trap shortly after starting the O_3–O_2 flow.

In the case of the alkylmercuric halides, the rate equation appears as,

$$\frac{-d[\mathrm{RHgX}]}{dt} = k\,[\mathrm{RHgX}]\,[\mathrm{O_3}] \qquad (14)$$

Since we may consider $k[O_3]$ to be constant after initial saturation is achieved, the integrated rate expression is that in Equation 15.

Table I. Products

Organomercurial	Reaction Temperature, °C
1. $CH_3CH_2CH_2HgBr$	10
2. $CH_3CH_2CH_2HgBr$	0
3. $(CH_3CH_2CH_2)_2Hg$	10
4. $(CH_3CH_2CH_2)_2Hg$	0
5. $(CH_3CH_2CH_2)_2Hg$	−76
6. $CH_3(CH_2)_5HgBr$	10
7. $CH_3(CH_2)_5HgBr$	10
8. $[CH_3(CH_2)_5]_2Hg$	10
9. $[CH_3(CH_2)_5]_2Hg$	0
10. $[CH_3(CH_2)_5]_2Hg$	−76
11. $(CH_3)_3CCH_2HgCl$	10
12. $(CH_3)_2CH(CH_2)_2HgBr$	10
13. $(CH_3)_2CHHgCl$	10

of Ozonolysis[a]

Inorganic Products	*Organic Products*
—	44% CH_3CH_2COOH 54% CH_3COOH
—	48% CH_3CH_2COOH 50% CH_3COOH
45% $HgCl_2$ 35% Hg_2Cl_2 20% HgO	64% CH_3CH_2COOH 20% CH_3COOH 15% $HCOOH$
—	67% CH_3CH_2COOH 19% CH_3COOH 14% $HCOOH$
—	72% CH_3CH_2COOH 16% CH_3COOH 12% $HCOOH$
—	30% $CH_3(CH_2)_4COOH$ 40% $CH_3(CH_2)_3COOH$ 10% $CH_3(CH_2)_2COOH$ 5% CH_3COOH
—	30% $CH_3(CH_2)_4COOH$ 40% $CH_3(CH_2)_3COOH$ 10% $CH_3(CH_2)_2COOH$ 5% CH_3COOH
—	50% $CH_3(CH_2)_4COOH$ 25% $CH_3(CH_2)_3COOH$ 10% $CH_3(CH_2)_2COOH$ 10% CH_3CH_2COOH 5% CH_3COOH
—	70% $CH_3(CH_2)_4COOH$ 10% $CH_3(CH_2)_3COOH$ 5% $CH_3(CH_2)_2COOH$ $<$5% CH_3CH_2COOH $<$5% CH_3COOH
—	80% $CH_3(CH_2)_4COOH$ 5% $CH_3(CH_2)_3COOH$ $<$5% other acids
—	— $(CH_3)_2CO$ — $(CH_3)_3COH$ — $(CH_3)_3CCOOH$
—	Complex mixture
77% $HgCl_2$ 23% Hg_2Cl_2	79% $(CH_3)_2CO$ 19% CH_3COOH

Table I.

Organomercurial	Reaction Temperature, °C
14. $[(CH_3)_2CH]_2Hg$	10
15. 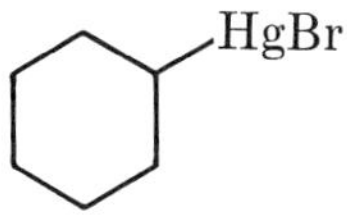	10
16. 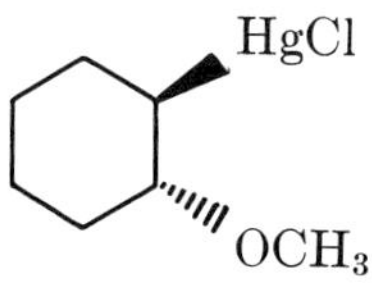	10
17. $(CH_3)_3CHgCl$	10

[a] Carrier gas was O_2 for all reactions except Reaction 7 where it was N_2.

$$-\int_{R_0}^{R} \frac{d[\mathrm{RHgX}]}{[\mathrm{RHgX}]} = k' \int_0^t dt \qquad (15)$$

where $k' = k\,[O_3]$

therefore $\ln\,[\mathrm{RHgX}] - \ln\,[\mathrm{RHgX}]_o = -k't$

or $\ln\,[\mathrm{RHgX}] = -k't + \ln\,[\mathrm{RHgX}]_o$

Least-squares plots of ln [RHgX] *vs.* t by an IBM 1620 computer gave values for the pseudo-first-order rate constant k'. Also, assuming a first-order rate dependence on ozone concentration, k was calculated *via* experimental saturation values for O_3 in $CHCl_3$ at 0°C. The results of these calculations are shown in Table III. From Table III a few relative rate sequences can be formulated. These are listed below.

Changing the mercury ligand:

$$\mathrm{RHgR} >> \mathrm{RHgI} > \mathrm{RHgBr} > \mathrm{RHgCl} \qquad (16)$$

Changing the R-group:

$$3° > \text{benzyl} > 2° > \text{methyl} > 1° \qquad (17)$$

Continued

Inorganic Products	*Organic Products*
55% $HgCl_2$ 35% Hg_2Cl_2 10% HgO	89% $(CH_3)_2CO$ 10% CH_3COOH
—	60% 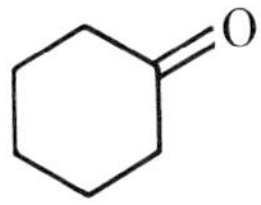 38% $HOOC(CH_2)_4COOH$
—	—[b] 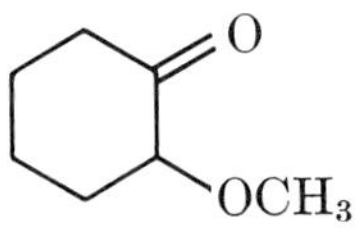
—	50% $(CH_3)_3COH$ 22% $(CH_3)_2CO$ 28% HCOOH

[b] This was the only organic product identified; yield not determined.

A second method of measuring the rate constants of ozonation was attempted. This alternate method involved the pre-saturation of a known volume of $CHCl_3$ at 0°C with ozone. An equimolar amount of organomercurial in $CHCl_3$ was then added to the O_3 solution, and both reagents were allowed to decrease in concentration. This attempted second-order rate analysis failed, however, since the necessarily low concentration of ozone (0.0163*M*) did not yield reliable NMR mercurial integrations. Hence the data were quite scattered and inconsistent.

Discussion

Ozonolysis Products. The ozonolysis of alkylmercuric halides and dialkylmercurials produced acids, ketones, and alcohols from primary, secondary, and tertiary organomercurials, respectively. The reactions also produced mercurous and mercuric salts in varying degrees. Although carbon–carbon scission frequently competed with the main carbon–mercury cleavage reaction, it is evident that the occurrence of the former reaction can be held to a minimum if dialkylmercurials are chosen as the reacting organomercurial and if low reaction temperatures are used. While the exact mechanism of the chain-shortening reaction is unknown, it was determined that neither the carboxylic acids nor ketones were the sources of the lower molecular weight products.

$$\begin{aligned} CH_3(CH_2)_4COOH + O_3,O_2 &\not\longrightarrow CH_3(CH_2)_3COOH \\ (CH_3)_2CO + O_3,O_2 &\not\longrightarrow CH_3COOH \end{aligned} \quad (18)$$

The possibility of carbon–carbon cleavage *via* CO_2 loss, while not proved conclusively, was at least suggested by the studies in Freon 114. Studies in this laboratory are now being done to prove the mechanism of this side reaction.

Intermediates. The NMR-observed formation of alcohols during the ozonolysis of secondary alkylmercuric halides, coupled with a rough rate profile for the appearance and/or disappearance of the mercurial, alcohol, and ketone, suggests that the initial ozonolysis product was largely alcohol.

$$R_2CHHgX \xrightarrow[CH_2Cl_2]{O_3,O_2} R_2CH{-}OH \xrightarrow{O_3,O_2} \text{ketone} \quad (19)$$

The exact percentage of ketone formation *via* the alcohol intermediate cannot be defined at this point. Moreover, no evidence exists for the formation of similar intermediates from the analogous primary mercurials. Further rate studies, including computer-assisted calculations for coupled consecutive and competing reaction systems, are clearly needed to solve this complex problem.

The need for further investigation into the dialkylmercurial intermediates is also indicated. Rough rate profiles for these reactions suggest that the majority of the initial organic product from cleavage of the first carbon–mercury bond was a carbonyl function.

$$\begin{aligned} (R_2CH)_2Hg \xrightarrow[CH_2Cl_2]{O_3,O_2} &R_2CO + R_2CHHgCl \\ &+ R_2CHOH \text{ (minor)} \end{aligned} \quad (20)$$

The simultaneous formation of a mole of alkylmercuric halide was quite surprising. However, these results parallel closely those obtained by oxygenation of dialkymercurials in chlorinated solvents (*18*). Razuvaev and co-workers got high yields of both isopropylmercuric chloride and acetone from O_2 treatment of diisopropylmercury in CCl_4 or $CHCl_3$ at 20°C (*18*). Obviously, a free radical intermediate of some type is suggested by these results.

Table II. Stoichiometry of Ozone Uptake

	Organomercurial	*Moles O_3/Hg cpd @0°C.*
1.	$(CH_3CH_2)_2Hg$	1.20[a]
2.	$(CH_3CH_2CH_2)_2Hg$	1.00[a]
3.	$[CH_3(CH_2)_5]_2Hg$	1.00[a]
4.	$[(CH_3)_3CCH_2]_2Hg$	1.09[a]
5.	$(\phi CH_2)_2Hg$	1.06[a]
6.	$[(CH_3)_2CH]_2Hg$	0.91[a]
7.	$[CH_3CH_2(CH_3)CH]_2Hg$	0.95[a]
8.	CH_3CH_2HgBr	1.31
9.	ϕCH_2HgCl	1.95
10.	$(CH_3)_2CHHgCl$	1.55, 1.85
11.	$(CH_3)_3CHgCl$	1.00

[a] These values are moles of O_3 per cleavage of the *first* C–Hg bond.

Reaction of Oxygen. The role of oxygen during the ozonolysis of alkylmercuric halides appears to be minor. Not only did oxygen fail to react with these mercurials at an appreciable rate, but substitution of nitrogen for oxygen as the ozone carrier gas failed to change either the yields or identity of the final products.

The dialkylmercurials present a different situation, however, as the literature reports that R_2Hg compounds do react with oxygen at an appreciable rate (*18, 19*). The reaction apparently causes a disproportionation of the dialkylmercurial (*18*).

$$[(CH_3)_2CH]_2Hg \xrightarrow{O_2} [(CH_3)_2CH]_2Hg \cdot O_2 \swarrow Hg^\circ + (CH_3)_2CHOH + (CH_3)_2CO \qquad (21)$$

Since the effect of oxygen on the products of dialkylmercury ozonations was not investigated in this study, the question of its role in these reactions remains in doubt. It should be clear that the entire yield of alcohol products from the reaction of di-*s*-alkylmercurials with ozone *might* be the result of the reaction of the organometallic with oxygen. A more detailed study of oxygen's role in these reactions will be included in future investigations.

Reaction Stoichiometry. The results of the stoichiometry study (*see* Table II) were varied but did suggest a 1:1::ozone: mercurial ratio for the cleavage of a carbon–mercury bond. Several problems were evident in such a study. Since mostly only the highest oxidation states for carbon were observed in the products of partial ozonation of 1°, 2°, and 3° organomercurials, it could be assumed that ozonolysis of the C–Hg bond involved the slowest step in the total reaction sequence. Hence, some ozone was probably used up as oxidizer for the original organic products of the C–Hg cleavage reaction. Also, as previously noted, methylene chloride did react at an appreciable rate with ozone to produce carbon dioxide among other possible products. For this reason also, the apparent amount of ozone used to cleave a mole of organomercurial was probably greater than the actual amount.

In view of the above-mentioned problems, the range of values for the data of Table II may be explained. The dialkylmercurials reacted much faster with ozone than did the corresponding alkylmercuric halides. As such, the starting material was completely gone before the concentration of alkylmercuric halide and alcohol had started to decrease perceptibly.

$$R_2Hg \xrightarrow[\text{fast}]{O_3,O_2} RHgX + ROH \xrightarrow{\text{slow}} \text{ketone} \quad (\text{RHgX} \xrightarrow{\text{slow}} \text{ketone}) \qquad (22)$$

A measure of reacted ozone to reacted dialkylmercurial during the first part of the reaction, therefore, yielded a consistent 1/1 ratio, as shown by Reactions *1–7* in Table II.

The alkylmercuric halides, on the other hand, reacted slowly enough with ozone to permit kinetic studies. While attempts were made to monitor ozone uptake during the first part of the reactions, it was clear that the primary and secondary alkylmercuric halides formed many ozone-reactive intermediates soon after the introduction of O_3/O_2. Hence, in the case of isopropylmercuric chloride (Reaction *10,* Table II), isopropyl alcohol was formed and had begun to react appreciably with ozone well before the mercurial had half reacted. By comparison, the reaction of *tert*-butylmercuric chloride was considerably faster than the primary and secondary alkylmercuric halides (Table III). In this case an excellent 1:1 correlation with ozone was noted throughout the majority of the reaction. Only at the end of this reaction, when other organic products began to appear (NMR), did the reaction mixture demand more than one equivalent of ozone.

Kinetic Studies. Although it was not possible to measure the ozonation rates for dialkylmercurials by the techniques described earlier, the alkylmercuric halides reacted slowly enough at 0°C that quite good rate plots were obtained. These plots, performed by an IBM 1620 computer and on concentration data obtained under pseudo-first-order reaction conditions, yielded the relative rate data listed in Table III. These data suggested the two general relative-rate sequences illustrated by sequences 16 and 17 (*see* p. 88).

Table III. Ozonolysis Rates

Reactant	$10^3\, k'$ (*min*$^{-1}$)	k[a] (*l/mole-min*)	*Relative*[b] *Rate*	*Correlation Coefficient*
1. $(CH_3)_3CCH_2HgCl$	<0.417	—	<0.15	—
2. CH_3CH_2HgCl	2.766	0.169	1.00	0.989
3. CH_3CH_2HgBr	3.757	0.231	1.36	0.962
4. CH_3HgCl	7.245	0.444	2.62	0.997
5. $CH_3CH_2(CH_3)CHHgCl$	14.65	0.899	5.29	0.995
6. $(CH_3)_2CHHgCl$	20.02	1.228	7.24	0.990
7. $(CH_3)_2CHHgBr$	22.27	1.366	8.05	0.982
8. $(CH_3)_2CHHgI$	28.11	1.724	10.16	0.999
9. $(CH_3)_2CHOH$	41.71	2.560	15.1	0.994
10. ϕCH_2HgCl	51.29	3.147	18.54	0.989
11. $(CH_3)_3CHgCl$	117.6	7.217	42.51	0.981

[a] Based on $[O_3] = 0.0163M$ for a saturated solution of O_3 in $CHCl_3$ at 0°C.
[b] Based on the rate for $CH_3CH_2HgCl = 1.00$.

Sequence 16 clearly demonstrates electrophilic attack by ozone in these reactions. As noted by Jensen and Rickborn (*2*), the rate of electrophilic cleavage of a carbon-metal bond increases as the polarization of that bond increases. This, in turn, is a direct consequence of the electronegativity of the second atom attached to mercury. The following representations, based on Pauling electronegativity values, illustrate this relationship.

$$\begin{array}{ccccc} \delta+ & & \delta+ & & \delta\delta- \\ C & — & Hg & — & F \\ 2.5 & & 1.9 & & 4.0 \end{array} \qquad \begin{array}{ccccc} \delta+ & & \delta+ & & \delta- \\ C & — & Hg & — & Cl \\ 2.5 & & 1.9 & & 3.0 \end{array} \qquad \begin{array}{ccccc} \delta+ & & \delta+ & & \delta- \\ C & — & Hg & — & Br \\ 2.5 & & 1.9 & & 2.8 \end{array} \qquad (23)$$

$$\begin{array}{ccccc} \delta- & & \delta\delta+ & & \delta- \\ C & — & Hg & — & I \\ 2.5 & & 1.9 & & 2.5 \end{array} \qquad \begin{array}{ccccc} \delta\delta- & & \delta\delta+ & & \delta\delta- \\ C & — & Hg & — & C \\ 2.5 & & 1.9 & & 2.5 \end{array}$$

Thus, not only should alkylmercuric iodides and dialkylmercurials have a more nucleophilic carbon compared with the others, but their respective –Hg–X functions should also be better leaving groups in electrophilic substitution reactions. This latter prophesy is supported in part by the results of Hughes *et al.* from their kinetic studies of mercury exchange reactions (*20*).

In transition state language, the fact that electrophilic substitution rates for organomercurials generally show the above trend (*i.e.*, RHgR > RHgI > RHgBr, etc.) (*2*) may be explained by assuming that a less energetic transition state exists when the developing positive charge on mercury is more fully dispersed. Hence, based on inductive and resonance effects, one would expect the transition state for C–HgR or C–HgI cleavage to be at a lower energy level than that for C–HgCl cleavage.

The above discussion assumes the usual inductive effect which halogens have on carbonium ions, but, of course, necessitates a reversal of the usual order for resonance effects. It is important to note that while fluorine can quite effectively form a pi bond with carbon, it cannot do so with the large mercury atom. Thus, structures such as 24 should be important only for the larger halogens.

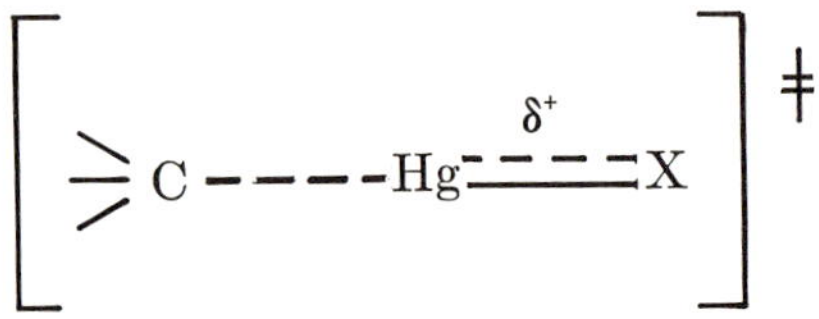

(24)

In this same regard, the transition state for cleavage of the dialkylmercurials must be of relatively lower energy, solely as a result of the inductive effect of the alkyl group since hyperconjugation effects should be nonexistent here.

Finally, in this study, as well as most others (*2*), substitution of carbon for halogen as the mercury ligand produced a huge rate increase for electrophilic cleavage. Although the Pauling electronegativity values for iodine and carbon are identical, the bonds of these two atoms to mercury are totally dissimilar. Hence, comparisons of these two systems should involve a complete treatment of orbital theory before one arrives at any conclusions concerning relative reactivities.

Turning to the other half of the kinetic study—that of rate effects from changes of the R group in RHgX—increased substitution on the carbon bound to mercury caused an appreciable reaction rate increase (Table III and sequence 17). These results certainly indicate the generation of partial positive charge on carbon in the transition state of the carbon-mercury cleavage reaction. While such a positive charge build-up might appear impossible at first consideration of the four general mechanisms for electrophilic cleavage (Reactions 3-6), minor modification of two of them adequately predicts it.

Jensen and Rickborn (*2*) discussed the possibility of transition states in which the electrophilic species was partially bound to the σ orbital of carbon–mercury bonds. Thus the fourth "center" of the four-center bromination mechanism was pictured as the carbon–mercury bond itself, or more specifically, the carbon sp^3 orbital (*2*).

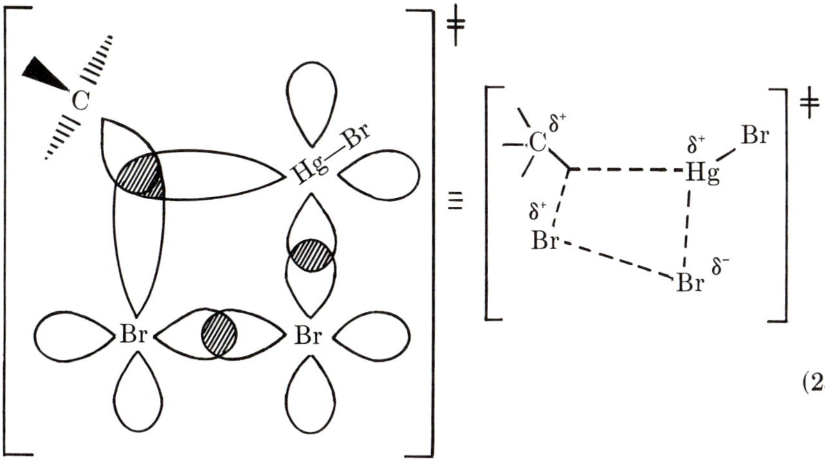

(25)

The carbon sp^3 atomic orbital was depicted as a solid line "indicating that its charge density is changed primarily by electronic factors rather than the actual making and breaking of the bond" (*2*). In a similar fashion the same authors illustrated the S_E2 cleavage of a dialkylmercurial by XHg^+.

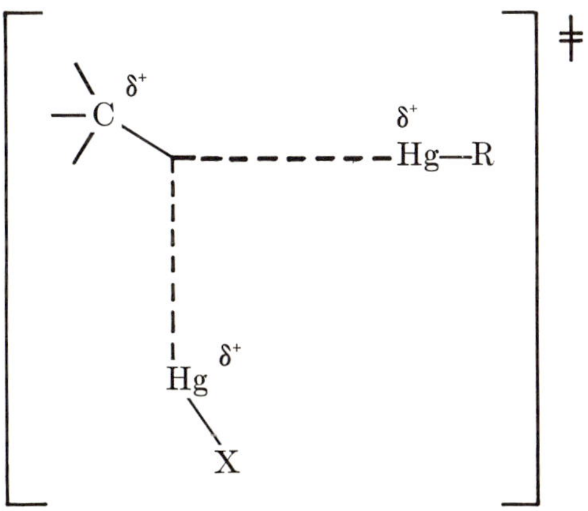

(26)

The authors stated that the positive charge would probably be divided between the three atomic centers and that electron-donating groups on carbon should increase the cleavage rate.

Mechanism. In consideration of a possible mechanism for cleavage of the carbon–mercury bond by ozone, it should be possible to rule out the S_E1 mechanism (Reaction 3) altogether. This mechanism clearly demands the formation of a carbanion in the rate-determining step. Likewise, the S_Ei mechanism for electrophilic cleavage seems improbable since the bimolecular rate-determining step involves a similar negative charge build-up on carbon (*4*). The transition state to such intermediates would obviously produce inverted relative-rate orders from that shown in sequence 17.

Of the two remaining mechanisms, the four-center and S_E2 processes, no selection is possible based on the data collected in this study. Moreover, since the S_E2 and S_Ei mechanisms are merely the two possible extreme versions of the four-center mechanism, the exact "purity" of an S_E2 or four-center process can often not be determined. One test which has been used to determine the particular involvement of the nucleophilic part of the attacking X-Y molecule during organomercurial cleavage reactions is to change the nature of Y such that it becomes a better or poorer nucleophile (*21*). If the reaction rate remains unchanged, the process is purely S_E2—*i.e.*, no interaction occurs between the Y group and mercury. This procedure of mechanistic determination is of course impossible for cleavage reagents such as ozone.

At this point in our studies we favor a five-membered cyclic transition state, quite analogous to the four-center transition state (Reaction 5), except that no σ bonds of the attacking electrophile are broken. The fourth bonding "center" is the carbon sp^3 orbital as in structures 25 and 26.

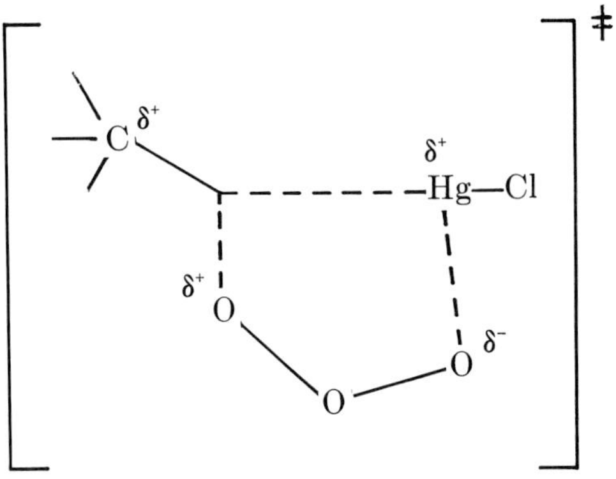

(27)

Although no kinetic evidence exists at this time to include the nucleophilic oxygen in such a bonding process (*i.e.*, S_E2 *vs.* four-center), it is difficult to believe that a negatively charged oxygen would not participate in the nearby formation of a positively charged mercury atom.

Such a five-membered cyclic transition state has been proposed for the triiodide cleavage of carbon-mercury bonds (*2*). While such a structure is unlikely for the linear I_3^- ion, it should easily accommodate the ozone molecule whose bonding angle is 116° (*12*). Of course the geometrical plausibility of Structure 27 depends on the degree of carbon–mercury bond breaking in the transition state. Models show that the extremely long carbon-mercury bond (2.06–2.20 A) (*22*) need not be stretched too much in order to permit a 1,3-dipolar attack by ozone as pictured.

The exact chronology of bond breaking and bond making is important in considering Structure 27 as the transition state for carbon–mercury cleavage. The initial attack by ozone must lie on the S_E2 side of the concerted, four-center process in order to allow for the build-up of some positive charge on carbon. Thus, the electrophilic oxygen should have achieved a greater degree of bond making with the carbon sp^3 orbital than that of the nucleophilic oxygen with the mercury atom by the time Structure 27 is reached. Attention is paid to this bond making order by the different lengths of partial bonds in Structure 27 (*2*).

The overall importance of transition state 27 is that rate sequences 16 and 17 are adequately explained. It is only this type of energy maximum which predicts partial positive charge on both carbon *and* mercury.

As an alternate transition state, one might envision the result of initial attack by the 1,2-dipolar resonance form of ozone *via* a four-membered cyclic transition state.

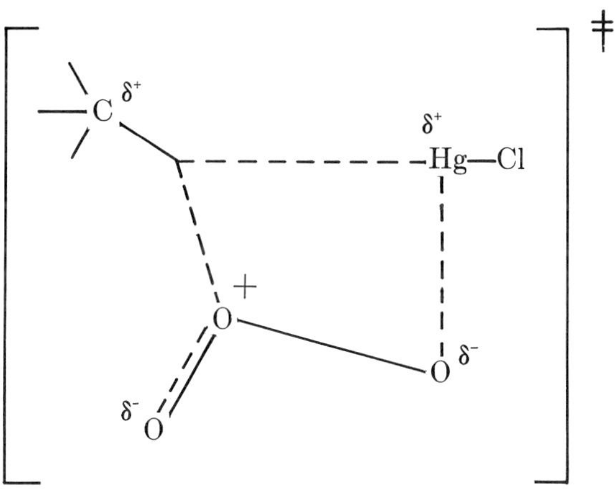

(28)

Since this transition state could yield an intermediate which in turn could conceivably produce the products observed, it cannot be ruled out at this point. Perhaps after more data have been collected concerning the existence of the analogous Staudinger intermediates (*23*) from olefin reactions, it will be possible to choose between the two reaction pathways.

Following the proposed transition state 27, a likely intermediate would be a trioxide similar to that proposed by Ouellette for silicon–hydrogen cleavage (Reaction 12).

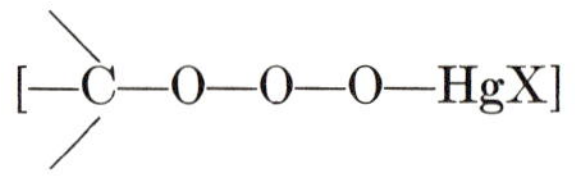
(29)

Although this intermediate was not isolated or even detected by NMR spectroscopy (at 31°C), its existence is inferred from the type of organic and inorganic products formed in the reaction. Future low temperature NMR studies of the ozonolysis solutions will be done to verify Structure 29.

The trioxide intermediate may decompose by many pathways. Results obtained in this paper suggest that these decomposition pathways might explain the two classes of mercury salts produced (Hg^{2+} and Hg^{+}) as well as the two divisions of organic products formed (with and without carbon–carbon scission). This possibility looks particularly attractive when the acetone/formic acid and $HgCl_2/Hg_2Cl_2$ ratios are compared in Reaction *13*, Table I. However, before extending speculation on this matter, several future experiments and theoretical calculations must be pursued. Among these will be an effort to measure quantitatively the amount and type (singlet or triplet) of oxygen generated by an O_3/N_2 ozonolysis. Also, theoretical consideration will be given to the metalo-trioxyalkanes in light of published data on the known dialkyltrioxide systems (*24, 25*).

Synthetic Utility. One goal of this research was to investigate new synthetic uses of organomercurials. Although the synthetic utility of this reaction has not been extensively investigated, products such as that from Reaction *16* in Table I do suggest a possible utility to organic chemists. Thus, the preparation of α-alkoxyketones should be possible by consecutive alkoxymercuration and ozonation reactions.

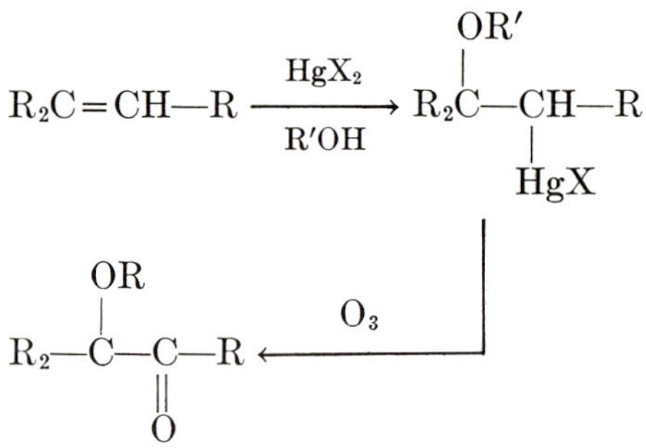

(30)

An attempt to prepare α-diketones *via* consecutive hydroxymercuration and ozonation failed, presumably because of the product's extreme reactivity towards ozone.

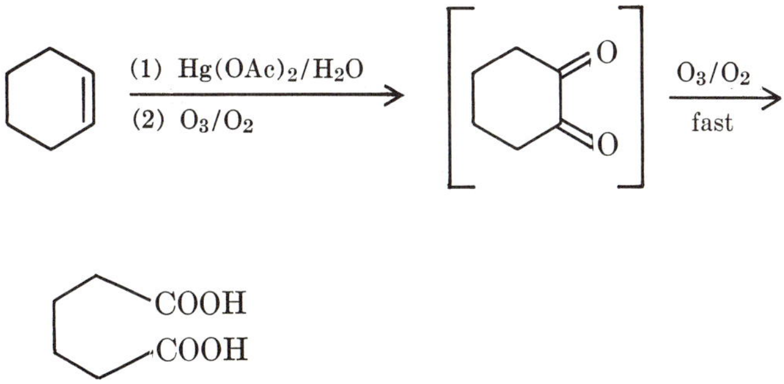

(31)

90%

A study is now underway to evaluate completely the future synthetic role of the oxymercuration–ozonolysis sequence.

Literature Cited

1. Jensen, F. R., Rickborn, B., "Electrophilic Substitution of Organomercurials," McGraw-Hill, New York, 1968.
2. Belew, J. S., "Oxidation," Vol. I, R. L. Augustine, Ed., p. 259, Marcel Dekkar, New York, 1969.
3. Aleksandrov, Yu. A., Sheyanov, N. G., Shushunov, V. A., *Zh. Obshch. Khim.* (1968) **38,** 1352–6; *CA* (1968) **69,** 77388q.
4. Aleksandrov, Yu. A., Sheyanov, N. G., *Zh. Obshch. Khim.* (1966) **36,** 953; *CA* (1966) **65,** 8955e.
5. Emel'yanov, B. V., Shemyakina, Z. N., Shvarov, V. N., *Khim. Prom.* (1968) **44,** 498–500; *CA* (1968) **69,** 7964y.
6. Aleksandrov, Yu. A., Sheyanov, N. G., *Zh. Obshch. Khim.* (1969) **39,** 141–3; *CA* (1969) **70,** 96847a.
7. Spialter, L., Austin, J. D., *Inorg. Chem.* (1966) **5,** 1975–8.
8. Ayrey, G., Barnard, D., Woodbridge, D. T., *J. Chem. Soc.* (1962) 2089–99.
9. Bockemuller, W., Pfeuffer, L., *Ann.* (1939) **537,** 178.
10. Ouellette, R. J., Marks, D. L., *J. Organometal. Chem.* (1968) **11,** 407–13.
11. Pike, P. E., Marsh, P. G., Erickson, R. E., Waters, W. L., *Tetrahedron Letters* (1970) 2679; Pike, P. E., MS Thesis, University of Montana (1970).
12. Rivera, J. T., MS Thesis, University of Montana (1971).
13. Marvel, C. S., Gauerke, C. G., Hill, E. L., *J. Amer. Chem. Soc.* (1925) **47,** 3009–11.
14. Gilman, H., Brown, R. E., *J. Amer. Chem. Soc.* (1929) **51,** 928.
15. Hatton, J. V., Schneider, W. G., Siebrand, W., *J. Chem. Phys.* (1963) **39,** 1330–6.
16. Kiefer, E. F., Waters, W. L., *J. Amer. Chem. Soc.* (1965) **87,** 4401.
17. Bailey, P. S., Reader, A. M., *Chem. Ind. (London)* (1961) 1063.
18. Ref. *2,* p. 98.
19. Brilkina, T. G., Shushunov, V. A., "Reactions of Organometallic Compounds with Oxygen and Peroxides," Chemical Rubber Co. Press, Cleveland, 1966.
20. Hughes, E. D., Ingold, C. K., Roberts, R. M. G., *J. Chem. Soc.* (1964) 3900.

21. Charmen, H. B., Hughes, E. D., Ingold, C. K., *J. Chem. Soc.* (1959) 2530.
22. Grdenic, D., *Quart. Rev.* (1965) **19,** 303–28.
23. Story, P. R., Alford, J. A., Ray, W. C., Burgess, J. R., "Abstracts of Papers," 161st National Meeting, ACS, March 1971, PETR A13.
24. Bartlett, P. D., Günther, P., *J. Amer. Chem. Soc.* (1966) **88,** 3288.
25. Bensen, S. W., *J. Amer. Chem. Soc.* (1964) **86,** 3922.

RECEIVED June 21, 1971. Supported by the National Science Foundation Grant No. GP-18317.

8

Ozonation of *p*-Nitro-*N*,*N*-Dimethylaniline

PER KOLSAKER

Universitetets Kjemiske Institutt, Blindern/Oslo 3, Norway

BRITA TEIGE

Odontologisk Institutt for Fysiologi og Biokjemi, Blindern/Oslo 3, Norway

The reaction of p-*nitro*-N,N-*dimethylaniline with ozone was studied. Only side-chain oxidation products,* p-*nitro*-N-*methylaniline,* p-*nitro*-N-*methylformanilide, and a dimeric peroxide, di-*[(N-*methyl*-N-p-*nitrophenyl)aminomethyl*] *peroxide, were formed. No* N-*oxide could be obtained. For comparison 2-carboxy-4-nitro*-N,N-*dimethylaniline,* p-*chloro*-N,N-*dimethylaniline, and* N,N-*dimethylaniline were ozonized. A mechanism is proposed.*

Ozonation of tertiary aliphatic amines has received some attention in recent years. Amine oxide formation was established long ago (*1*). Oxidation of alkyl groups was reported by Henbest and Stratford (*2*), Shulman (*3*), and Bailey *et al.* (*4, 5*). Formation of amine hydrochlorides was established when chlorinated solvents were used (*2–5*). Dealkylated products were formed (*2, 4, 5*), and amides were found by the same authors (*2–5*).

In a previous paper (*6*) we reported on the ozonation of 2-carboxy-4-nitro-*N*,*N*-dimethylaniline (I) and *p*-nitro-*N*,*N*-dimethylaniline (IVc). *N*-oxides were not found in these cases, and the products formed indicated only side-chain oxidation. The intermediacy of a carbinolamine II is indicated by the formation of the lactone III.

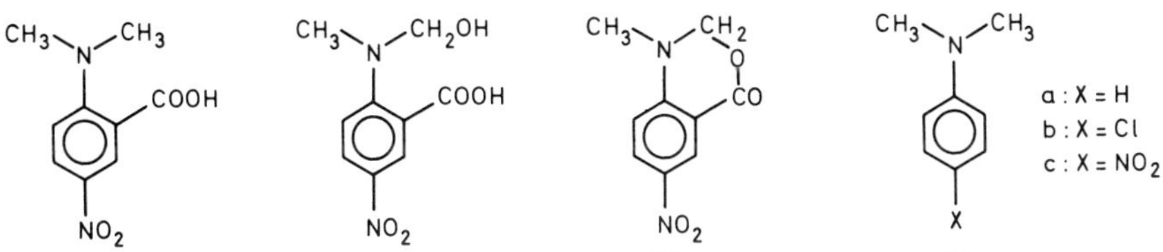

I II III IV

Table I. Ozonation of 2-Carboxy-4-nitro-*N,N*-dimethylaniline (I)

Solvent	*Demethylated Starting Material, %*	*Lactone (III), %*
CH_2Cl_2	81	18
EtOAc	71	28
MeOH	64	32

Demethylated starting material was also a product of the ozonation. Product composition varied with solvent as shown in Table I.

When *p*-nitro-*N,N*-dimethylaniline (IVc) was ozonized at 0°C in ethyl acetate, methylene chloride, or methanol, a mixture of products resulted. In addition to the expected side-chain oxidation products—*p*-nitro-*N*-methylaniline (Vc), *p*-nitro-*N*-methyl formanilide (VIc)—a peroxide compound was formed. This peroxide, which is not formed until the solvents are removed, was shown by a series of experiments (described below) to be identical with di-[(*N*-methyl-*p*-nitrophenyl)-aminomethyl]peroxide (VII). Deoxygenation of VII with triethyl phosphite (7) yielded the ether VIII, which in turn decomposed at its melting point to the amine IX.

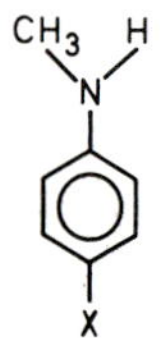

V

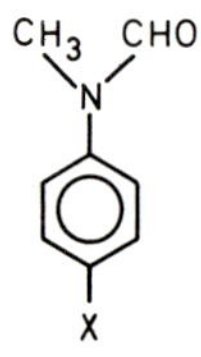

VI

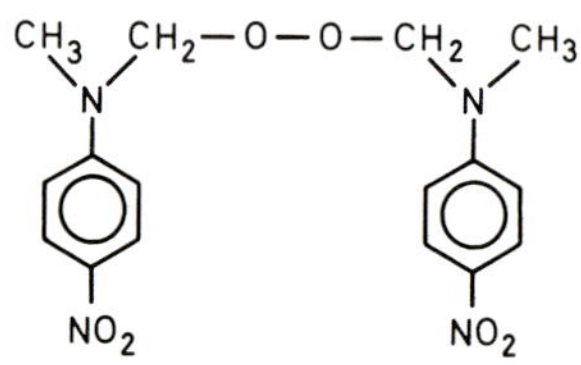

VII

a : X = H , b : X = Cl , c : X = NO_2

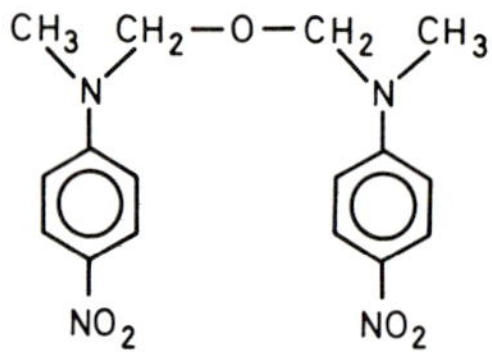

VIII

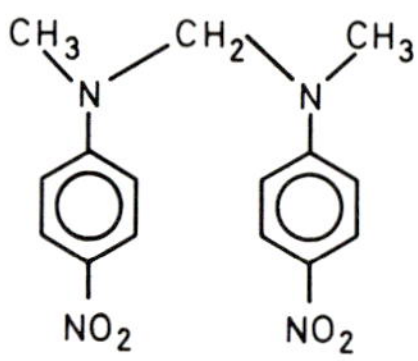

IX

We now report our results when the experimental conditions are systematically varied. Ozonations of some other *N,N*-dimethylanilines are also included. The results are given in Table II.

Experimental

General. The ozonation procedures are those described in the literature, both using ozone–oxygen and ozone–nitrogen. Oxygen analyses were performed using a Beckman F3 oxygen analyzer as described previously (*8*). Gas liquid chromatographic (GLC) determinations were carried out with an Aerograph model A 700, using an 8.5-ft × 0.25-inch column of polyethylene glycol 20*M* (30% on Chromosorb WHMDS, 60/80 mesh) for Experiments 1 and 2.

Proton magnetic resonance (PMR) spectra were recorded on a Varian A 60A spectrometer using ordinary chloroform as solvent and tetramethylsilane as internal reference. Melting points were not corrected.

Materials. 2-Carboxy-4-nitro-*N*,*N*-dimethylaniline (I) was synthesized from its chloro analog with dimethylamine (aqueous solution), mp 164°–166°C. *p*-Chloro-*N*,*N*-dimethylaniline (IVb) and *N*,*N*-dimethylaniline (IVa) were commercially available in the pure state and were used as supplied. *p*-Nitro-*N*,*N*-dimethylaniline (IVc) was prepared from *p*-nitrochlorobenzene and dimethylamine (*9*); mp 166°C.

Ozonation of 2-Carboxy-4-nitro-*N*,*N*-dimethylaniline (I). The amine (2.10 grams, 10 mmoles) was dissolved in the appropriate solvents (200 ml methanol, ethyl acetate, or methylene chloride) and ozonized at 0°C with equimolar amounts of ozone; quantitative absorption occurred. Near the end of the ozonation time, a precipitate formed. In ozonations in methanol this substance was identical with lactone III; mp 176°–177°C, prepared according to Villiger (*10*). In the other solvents the insoluble product was the demethylated starting material, mp 265°–268°C (dec.); reported mp 263°–264°C (*11*). The mother liquor was evaporated, and the two compounds were separated by fractional crystallization. The results are shown in Table I.

Ozonation of *N*,*N*-dimethylaniline (IVa). The amine (10 mmoles) was dissolved in methylene chloride (150 ml) and ozonized at −78°C with 10 mmoles of ozone (nitrogen as carrier gas) (*12*); quantitative absorption occurred. The solution became dark brown, and finally a tarry substance separated out. This substance, which adhered to the walls of the ozonation vessel, was repeatedly digested with methylene chloride. The combined methylene chloride extracts, after washing with water, were concentrated and analyzed by GLC. The results are shown in Table II.

Ozonation of *p*-Chloro-*N*,*N*-dimethylaniline (IVb). The amine (10 mmoles) was dissolved in methylene chloride (100 ml) and ozonized at −78°C with 10 mmoles of ozone. Quantitative absorption occurred. The solution was carefully washed with water and then analyzed by GLC. For the results *see* Table II.

Ozonation of *p*-Nitro-*N*,*N*-dimethylaniline (IVc). Concentration, temperature, and solvents were varied as shown in Table II. In a typical run, 5 mmoles of amine were dissolved in 200 ml of ethyl acetate and ozonized at 0°C with 5 mmoles ozone, using nitrogen as carrier gas; quantitative absorption occurred. The exit gas was analyzed for molecular oxygen. After the solution had come to room temperature, it was still clear and homogeneous. The solvent was evaporated, and the residue was dried and weighed. Ethyl acetate (200 ml) was added. A yellow compound (VII) was now insoluble and was removed by filtration. The filtrate was evaporated, and the residue was examined by PMR. The amounts of *p*-nitro-*N*,*N*-dimethylaniline (IVc, starting material), *p*-nitro-*N*-methylaniline (V), and *p*-nitro-*N*-methyl formanilide (VI) were esti-

Table II. Ozonation

Expt.	*Compound*	*Amine, mmoles*	*O_3/O_2 mmoles*	*O_3/N_2 mmoles*	*Solvent (ml)*	*Temp., °C*
1	IV a	10		10	CH_2Cl_2(150)	−78
2	IV b	10	10		CH_2Cl_2(100)	−78
3	IV c	10		10	EtOAC(200)	−78
4	IV c	5		5	EtOAc(200)	−78
5	IV c	5		5	MeOAc(200)	−78
6	IV c	5		5	CH_2Cl_2(200)	−78
7	IV c	10		10	EtOAc(200)	−40
8	IV c	10	10		EtOAc(100)	−40
9	IV c	5		5	EtOAc(200)	−40
10	IV c	5		5	MeOAc(200)	−40
11	IV c	5		5	CH_2Cl_2(200)	−40
12	IV c	10		10	EtOAc(100)	0
13	IV c	10		10	EtOAc(200)	0
14	IV c	5	5		EtOAc(200)	0
15	IV c	5		5	EtOAc(200)	0
16	IV c	5		5	MeOAc(200)	0
17	IV c	5		5	CH_2Cl_2(200)	0
18	IV c	10	10		EtOAc(200)	25
19	IV c	5		5	EtOAc(200)	25
20	IV c	5		5	MeOAc(200)	25
21	IV c	5		5	CH_2Cl_2(200)	25
22	IV c	5		10	EtOAc(200)	−78
23	IV c	5		10	EtOAc(200)	−40
24	IV c	5		10	EtOAc(200)	0
25	IV c	5		10	EtOAc(200)	25
26	IV c	5		2.5	EtOAc(200)	−78
27	IV c	5		2.5	EtOAc(200)	−40
28	IV c	5		2.5	EtOAc(200)	0
29	IV c	5		2.5	EtOAc(200)	25

mated by measuring the integrated areas for the *N*-methyl protons. The limit of error in such a determination is probably around 5%.

Insoluble Compound VII. This compound melted at 176°–177°C (acetonitrile–water). It was identical (by infrared) with the product obtained when *p*-nitro-*N*-methylaniline, formaldehyde, and hydrogen peroxide were heated in ethanol (7).

Active Oxygen Content. A total of 181.2 mg (0.5 mmole) of VII were dissolved in 75 ml of acetonitrile; 1 gram of sodium iodide in 25 ml acetonitrile was added, and nitrogen was passed through the solution. No reaction occurred; when 2 ml perchloric acid (60%) were added, the

Experiments

O_2, *mmoles*	*V*, *mmoles* (%)	*VI*, *mmoles* (%)	*VII*, *mmoles* (%)	*Rec. IV*, *mmoles* (%)	*Remarks*
	Products				
7.9	1.0(10)	2.0(20)	0	2.4(24)	tar formation
—	1.7(17)	2.6(26)	0	2.0(20)	tar formation
6.5	0	1.3(13)	0	8.7(87)	
4.0–4.7	0	0.5–1.0 (10–20)	0	4.0–4.5 (80–90)	several runs
4.4	0	1.3(26)	0.1(4)	3.5(70)	
4.5	2.6(52)	1.2(24)	0.6(24)	0	
6.9	1.7(17)	2.5(25)	1.6(32)	2.7(27)	
—	0.6(6)	2.7(27)	1.5(30)	4.4(44)	
4.1	0.7(14)	1.8(36)	0.8(32)	0.8(16)	
4.1	0.9(18)	1.7(34)	0.7(28)	0.9(18)	
3.9	2.1(42)	1.2(24)	0.7(28)	0.2(4)	
n.c.	0.8(8)	2.7(27)	1.9(38)	2.7(27)	no oxygen analysis
9.0	1.1(11)	3.3(33)	2.2(44)	1.3(13)	
—	1.0(20)	1.4(28)	0.9(36)	0.8(16)	
4.1	0.8(16)	1.5(30)	1.0(40)	0.7(14)	
4.0	0.6(12)	1.6(32)	0.9(36)	0.9(18)	
3.4	1.5(30)	1.2(24)	0.7(28)	0.7(14)	
—	0.3(3)	3.5(35)	2.2(44)	1.6(16)	
4.0	0	1.8(36)	1.0(40)	1.2(24)	
4.3	traces	1.2(24)	0.7(28)	2.2(44)	
3.3	1.3(26)	1.3(26)	0.8(32)	0.8(16)	
7.9	0	2.3(46)	0	2.5(50)	0.3 mmoles diperoxide (12%)
8.4	0	3.3(66)	0	traces	0.9 mmoles diperoxide (36%)
6.1	0	3.7(74)	0	0	0.7 mmoles diperoxide (28%)
6.9	0	4.3(86)	0	0	0.4 mmoles diperoxide (16%)
2.1	0	0	0	5.0(100)	
1.9	0	0.6(12)	0.4(16)	3.6(72)	
1.5	0	0.6(12)	0.5(20)	3.3(66)	
1.3	0	0.5(10)	0.3(12)	3.9(78)	

brown color of iodine appeared immediately. After 20 minutes at room temperature the solution was titrated with 0.1*N* sodium thiosulfate. The endpoint was difficult to observe, but approximately 10–11 ml of titer solution were used, corresponding to about 100% active oxygen. Solid sodium bicarbonate was then added, and the acetonitrile was evaporated. Yellow crystals were filtered off, melting at 150°–152°C. Infrared spectra (KBr pellet) confirmed the identity of the substance as *p*-nitro-*N*-methylaniline (V); the yield was 133 mg (87.5%).

THERMAL DECOMPOSITION: Compound VII (101.7 mg) was heated neat at 155°C (boiling anisole). An infrared spectrum (KBr pellet) of the residue showed it to be identical with *p*-nitro-*N*-methylformanilide (VI); the yield was 96 mg (95%).

Reaction with Triethyl Phosphite (*7*). Refluxing VII in triethyl phosphite yielded a very insoluble compound (VIII) melting at 177°–178°C. (Found: C 55.4, H 5.3, N 26.4. Calc. ($C_{16}H_{18}N_4O_5$): C 55.5, H 5.2, N 16.2).

Decomposition of VIII. Heating VIII neat at its melting point yielded IX; mp 276°–278°C (Found: C 56.9, H 5.1, N 17.3. Calc. ($C_{15}H_{16}N_4O_4$): C 57.0, H 5.1, N 17.7). Compound IX could also be synthesized from *p*-nitro-*N*-methylaniline and formaldehyde.

Ozonation of *p*-Nitro-*N*,*N*-dimethylaniline with Two Mole Equivalents of Ozone. The products from this ozonation were obtained and analyzed by the same procedure as for the equimolar runs. The insoluble compound (XVIII) melted at 150°C. (Found: C 49.0, H 4.4, N 14.3. Calc. ($C_{16}H_{16}N_4O_8$): C 49.0, H 4.1, N 14.3).

Results

(1) There seems to be no significant difference in product yields using either ozone–oxygen or ozone–nitrogen (*e.g.*, Experiments 14 and 15). The small differences may be the result of different rates of ozone delivery.

(2) Use of methylene chloride as solvent consistently leads to higher yields of demethylated product (V), accompanied mostly by a decrease in recovered starting material. Lower oxygen yields are also observed at all temperatures except at −78°C.

(3) Ozone appears to be decomposed by *p*-nitro-*N*,*N*-dimethylaniline in ethyl and methyl acetate at −78°C (*cf.* Experiments 3, 4, 5, and especially 26).

(4) The only temperature effect of any significance is observed in ethyl or methyl acetate in going from −40° to −80°C (Experiments 4 and 9).

(5) The higher yields of the formanilide VI are accompanied by higher yields of the peroxide VII.

(6) Since starting material always is recovered in the equimolar runs, the initial attack of ozone must be somewhat slower than one (or several) of the subsequent steps involving ozone and the primary reaction product(s).

Discussion

Several authors (*1, 3, 4*) propose that ozone attacks amines electrophilically at the nitrogen lone pair. Additional support for this comes from the following observation. Ozone passes unchanged through solution of *p*-nitro-*N*,*N*-dimethylaniline (IVc) [or *N*,*N*-dimethylaniline (IVa)] in 70% perchloric acid. Diluting the acid to about 20–25% produces the yellow color of unprotonated amine, and ozone is now absorbed quantitatively. Since the yellow color of the free amine (IVc) disappears in concentrated acid solution, the amine nitrogen (and not another point of the molecule) must be protonated.

NMR studies show that amides are protonated at the oxygen atom, indicating a lower electron density at the nitrogen atom in amides than even in *p*-nitro-*N*,*N*-dimethylaniline. About 50% excess ozone had to be passed through a solution of *N*,*N*-dimethylbenzamide and as much as 125% excess through *p*-nitro-*N*,*N*-dimethylbenzamide to ensure a complete reaction.

Since the products of the ozonation of IV only indicate side-chain oxidation, it is possible that ozone insertion in the carbon–hydrogen bond is the starting point of the mechanism. Again the answer may be found in NMR spectroscopy, bearing in mind that the various resonance positions reflect the electron density near the hydrogen atom in a carbon–hydrogen bond (in the absence of strong anisotropic groupings). The chemical shift difference for the methyl protons in going from *N*,*N*-dimethylaniline to its conjugate acid is 0.45 ppm, indicating a somewhat lower electron density in the carbon–hydrogen bond. The chemical shift difference for the methylene group protons going from triethylamine to diethyl ether is about 1 ppm (*12*). For both these compounds side-chain oxidation takes place although a much larger electron density change than in amine–conjugate acid pair is indicated by NMR. Thus, the inability of ozone to react with the conjugate acid must be linked to the absence of the lone pair on the amine nitrogen.

From these considerations it seems safe to conclude that the *initial* attack of ozone on amines takes place at the nitrogen atom. This is also logical since the electrophilic ozone molecule must interact with the center of highest nucleophilicity in the substrate—*viz.*, the lone pair of the nitrogen atom.

The secondary reactions will then be determined by the rest of the molecule. Whether amine oxide formation will be competitive with side-chain oxidation or not depends on the strength of the nitrogen–oxygen bond in the primary adduct. This again is determined mainly by the electron density at the nitrogen atom of the substrate—*viz.*, the strength of the base. Aliphatic amines are stronger bases than the aromatic (trimethylamine, $pK_b = 4.2$, *N*,*N*-dimethylaniline, $pK_b = 9.6$). Amine oxides are formed from aliphatic amines, indicating that the bond between nitrogen and oxygen in the primary adduct is covalent and stronger than the oxygen–oxygen bond, which breaks down to form amine oxide and molecular oxygen. In the aromatic amines studied here the nitrogen–oxygen bond is weaker than a normal covalent bond and probably is more like a charge transfer bond, and secondary attack on a carbon–hydrogen bond of the side chain takes place as indicated in Reactions 1 and 3.

Table III. Ozonation Experiments

Expt.	*Solvent*	*Temp., °C*	*Consumed*[a] *IV*		*Consumed*[a] O_3		*Produced*[a, b] O_2	
			Calcd.	*Found*	*Calcd.*	*Found*	*Calcd.*	*Found*
1	CH_2Cl_2	−78	3.0	7.6	5.0	10.0	5.0	7.9
2	CH_2Cl_2	−78	4.3	8.0	6.9	10.0	6.9	
3	EtOAc	−78	1.3	1.3	2.6	10.0	2.6	6.5
4	EtOAc	−78	0.5–1.0	0.5–1.0	1.0–2.0	5.0	1.0–2.0	4.0–4.7
5	MeOAc	−78	1.5	1.5	2.7	5.0	2.7	4.4
6	CH_2Cl_2	−78	5.0	5.0	5.6	5.0	5.0	4.5
7	EtOAc	−40	7.4	7.3	8.3	10.0	6.7	6.9
8	EtOAc	−40	6.3	5.6	7.5	10.0	6.0	
9	EtOAc	−40	4.1	4.2	5.1	5.0	4.3	4.1
10	MeOAc	−40	4.0	4.1	5.0	5.0	4.3	4.1
11	CH_2Cl_2	−40	4.7	4.8	5.2	5.0	4.5	3.9
12	EtOAc	0	7.3	7.3	8.1	10.0	6.2	
13	EtOAc	0	8.8	8.7	9.9	10.0	7.7	9.0
14	EtOAc	0	4.2	4.2	4.7	5.0	3.8	
15	EtOAc	0	4.3	4.3	4.8	5.0	3.8	4.1
16	MeOAc	0	4.0	4.1	4.7	5.0	3.8	4.0
17	CH_2Cl_2	0	4.1	4.3	4.6	5.0	3.9	3.4
18	EtOAc	25	8.2	8.4	9.5	10.0	7.3	
19	EtOAc	25	3.8	3.8	4.6	5.0	3.6	4.0
20	MeOAc	25	2.6	2.8	3.1	5.0	2.4	4.3
21	CH_2Cl_2	25	4.2	4.2	4.7	5.0	3.9	3.3
22	EtOAc	−78	2.9	2.5	5.2	10.0	4.6	7.9
23	EtOAc	−40	5.1	5.0	8.4	10.0	6.6	8.4
24	EtOAc	0	5.1	5.0	8.8	10.0	7.4	6.1
25	EtOAc	25	5.1	5.0	9.4	10.0	8.4	6.9
26	EtOAc	−78	0	0	0	2.5	0	2.1
27	EtOAc	−40	1.4	1.4	1.6	2.5	1.2	1.9
28	EtOAc	0	1.6	1.7	1.7	2.5	1.2	1.5
29	EtOAc	25	1.1	1.1	1.3	2.5	1.0	1.3

[a] Yields are given in mmoles.
[b] No entry in column means value not checked.

Attack other than on the amino group apparently does represent a competitive reaction course when the substituents on the aromatic ring are less electronegative than the nitro group. This is demonstrated by the strong coloration and tar formation observed when *N,N*-dimethylaniline and *p*-chloro-*N,N*-dimethylaniline are ozonized and by the fact that more ozone is consumed by these compounds than can be accounted for by side chain oxidation (Table III, Experiments 1 and 2).

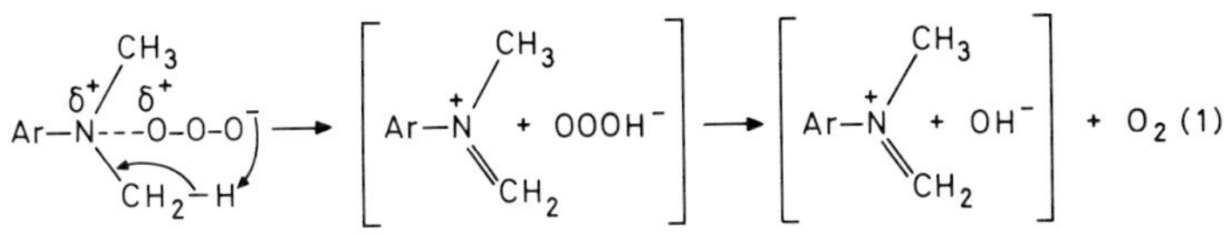

X XI XII

Attack of a second mole of ozone on XIII, similar to Reaction 1, will then lead to the *N*-methylformanilides VI (Reaction 3):

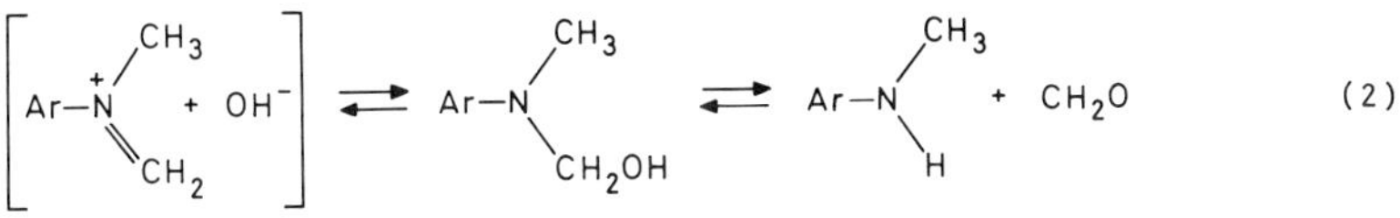

XII XIII V

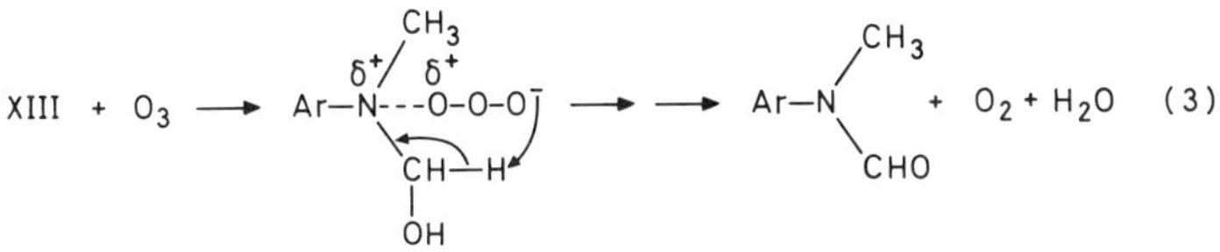

XIV VI

According to Rieche *et al.* (*13*) aliphatic carbinolamines may react with hydroperoxides to form α-aminoalkylperoxides. Horner and Knapp (*7*) prepared di-α-aminoalkylperoxides from *N*-alkylanilines, formaldehyde, and hydrogen peroxides. If the reaction sequences leading to these peroxides are those outlined below, a similar α-aminohydroperoxide (XV) probably also is the precursor in the formation of the di-α-aminoalkylperoxide obtained in the ozonation of *p*-nitro-*N*,*N*-dimethylaniline.

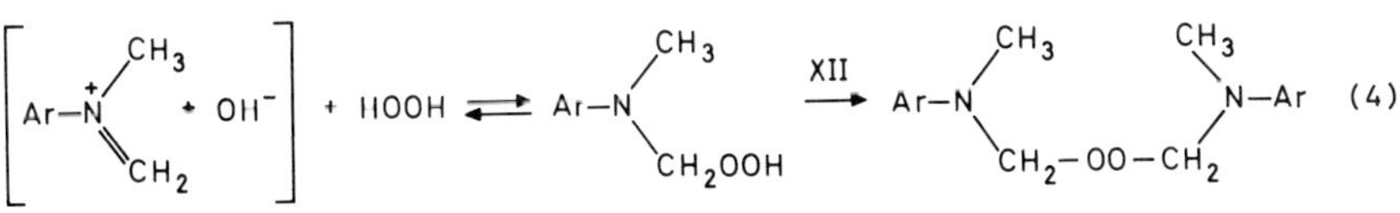

XII XV VII

The problem then is the route to hydrogen peroxide in the present case. Insertion of ozone into a carbon–hydrogen bond has been postulated by White and Bailey (*14*) to explain the ozonation of benzaldehydes, by Price and Tumolo (*15*) in the ozonation studies of ethers, and by Batterbee and Bailey (*16*) in their studies on the ozonation of anthrone. We postulate a similar intermediate to explain the formation of hydrogen peroxide and subsequently the diaminoperoxide VII.

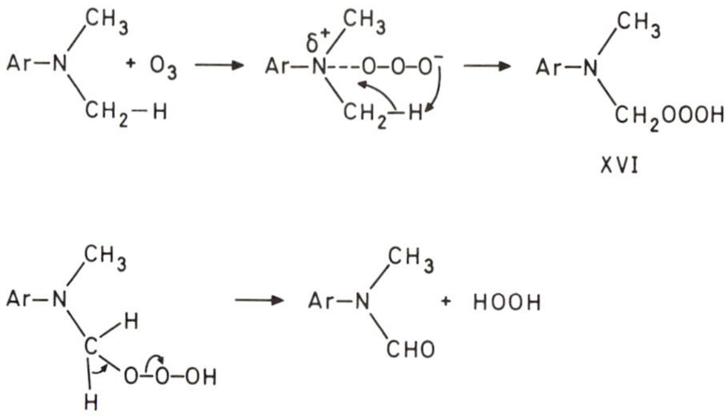

(5)

XVI VI

As seen above, *N*-methylformanilide is also produced by this route. The reaction routes pertinent to our studies are summarized in Scheme 1.

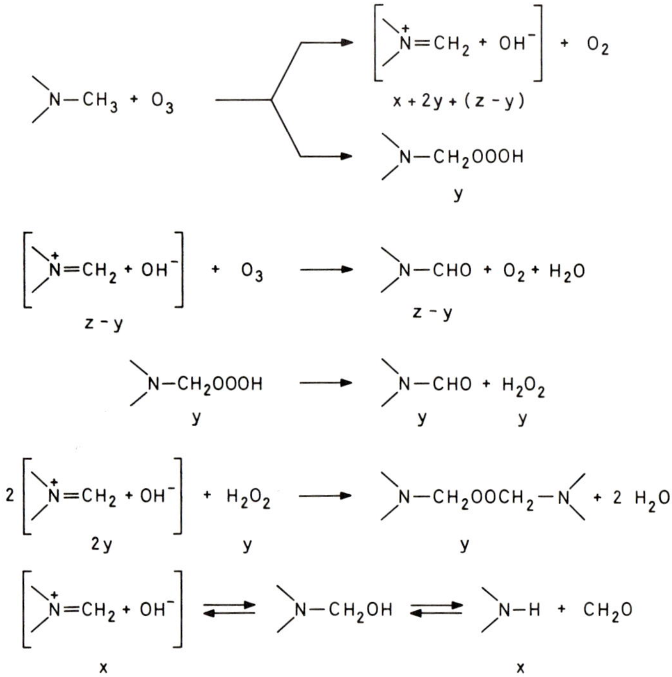

Scheme 1

From these sequences the following can be obtained:

Consumed starting material: $a = x + y + 2y + (z - y) = x + 2y + z,$

Consumed ozone: $b = x + 2y + z + (z - y) = x + y + 2z,$

Production of oxygen: $c = x + 2y + (z - y) + (z - y) = x + 2z,$

Here x is the yield of demthylated starting material (V), y is the yield of the diaminoperoxide VII, z is the yield of *N*-methylformanilides (VI). A comparison of the calculated values for a, b, and c with those obtained in the experiments, as seen from Table II, are shown in Table III. When 2 mole equivalents of ozone are used, a compound which apparently is a diperoxide is formed (*see* Table II). Elemental analysis indicates a molecular formula of $C_{16}H_{16}N_4O_8$, and acid hydrolysis gives small amounts of *p*-nitro-*N*-methylaniline (Vc). Formation of this diperoxide can be rationalized as follows:

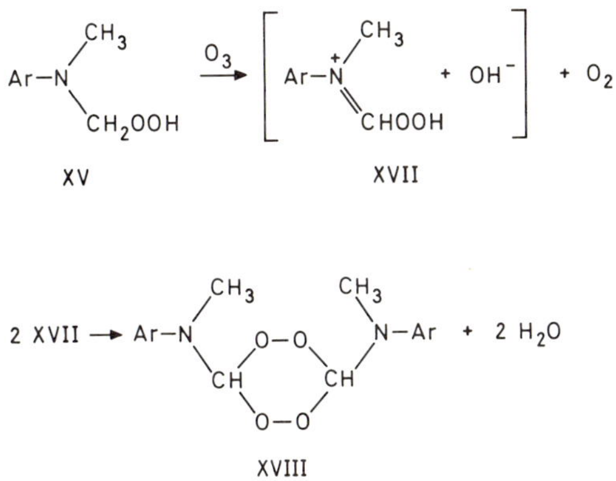

(6)

The entity XVII of questionable stability is thought to dimerize to the diperoxide XVIII. A modified reaction must then be applied to Experiments 22-25 (Scheme 2) (p. 112).

From Scheme 2:

Consumed starting material: $a = z + 2y'$

Consumed ozone: $b = (z + 2y') + (z - 2y') + 2y' = 2z + 2y'$

Production of oxygen: $c = z + (z - 2y') + 2y' = 2z,$

The value of y' is now the yield of dimeric diperoxide XVIII. The results are entered in Table III, Experiments 22-25.

Table III shows that the starting material is accounted for well. In Experiments 1 and 2 considerable attack on the aromatic nucleus must have taken place since much tar is formed in these ozonations. When *p*-chloro-*N*,*N*-dimethylaniline (IVb) is ozonized, more attack takes place

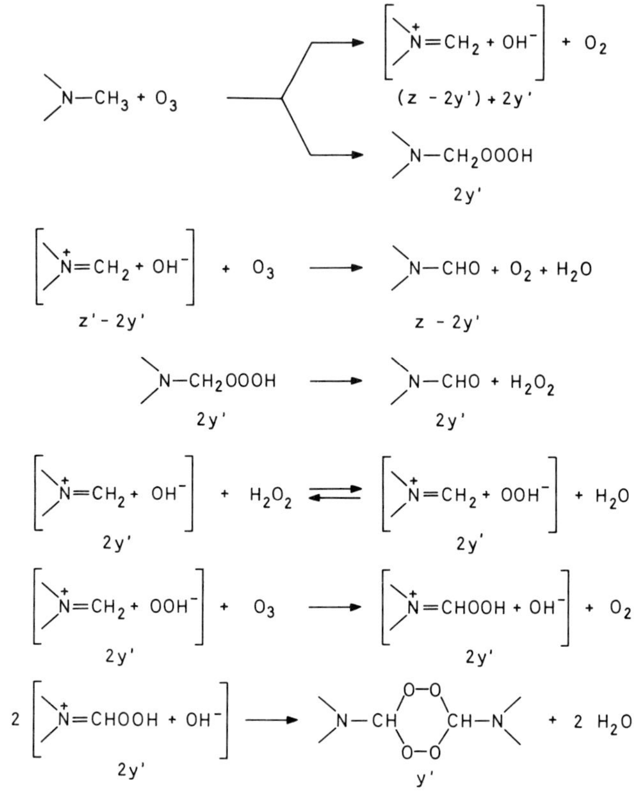

Scheme 2

on the amine function than is the case with the unsubstituted *N,N*-dimethylaniline (IVa). This is expected since the chlorine atom deactivates the ring for electrophilic attack. This trend is accentuated in going to the nitrosubstituted dimethylaniline (IVc), where ring attack is not observed.

The amount of ozone used for ring attack is not accounted for in our expressions; therefore the values for *b* and *c* in Experiments 1 and 2 will be erroneous. The large discrepancy in consumed ozone and ozygen yield when ozonizing in ethyl or methyl acetate at −78°C (Experiments 3-5, 22, and 26) cannot be accounted for by any of our schemes. The possibility of ozone attack on the solvents cannot be excluded. However, we have not been able to discover any product arising from oxidation of the solvent. Ozonation of ethyl acetate was shown by Price and Tumolo to be nonquantitative while ozone is quantitatively absorbed at −78°C when *p*-nitro-*N,N*-dimethylaniline (IVc) is dissolved in ethyl acetate. This problem has to be left open at this stage of our research.

Literature Cited

1. Bailey, P. S., *Chem. Rev.* (1958) **58,** 925.
2. Henbest, H. B., Stratford, J. W., *J. Chem. Soc.* (1964) 711.
3. Shulman, G. P., *Can. J. Chem.* (1965) **43,** 3069.
4. Bailey, P. S., Mitchard, D. A., Khashab, A.-I. Y., *J. Org. Chem.* (1968) **33,** 2675.
5. Bailey, P. S., Keller, J. E., Mitchard, D. A., White, H. M., ADVAN. CHEM. SER. (1968) **77,** 58.
6. Kolsaker, P., Meth-Cohn, O., *Chem. Commun.* (1965) 423.
7. Horner, L., Knapp, K. H., *Ann.* (1959) **622,** 79.
8. Bailey, P. S., Reader, A. M., *Chem. Ind.* (1961) 1063.
9. Shirley, D. A., "Preparations of Organic Intermediates," p. 221, Wiley, New York, 1951.
10. Villiger, V., *Chem. Ber.* (1909) **42,** 3534.
11. Morley, J. S., Simpson, J. C. E., *J. Chem. Soc.* (1948) 360.
12. Spiesecke, H., Schneider, W. G., *J. Chem. Phys.* (1961) **35,** 722.
13. Rieche, A., Schmitz, E., Beyer, E., *Chem. Ber.* (1959) **92,** 1206.
14. White, H. M., Bailey, P. S., *J. Org. Chem.* (1965) **30,** 3037.
15. Price, C. C., Tumolo, A. L., *J Amer. Chem. Soc.* (1964) **86,** 4691.
16. Batterbee, J. E., Bailey, P. S., *J. Org. Chem.* (1967) **32,** 3899.

RECEIVED June 7, 1971. Work supported by a grant from Det Videnskapelige Forskningsfond av 1919.

9

Mechanism of Ozonation Reactions

V. Feist's Ester

RONALD E. ERICKSON and GARY D. MERCER

University of Montana, Missoula, Mont. 59801

The ozonation of Feist's ester yields three products: 2,3-dicarbomethoxytetrahydrofuran-4-one (III), 3,4-dicarbomethoxy-5-hydroxy-2-oxa-2,3-dihydropyran (IV), and 4,5-dicarbomethoxy-1-oxaspiro[2.2]pentane (V). Structures were assigned primarily through spectroscopic evidence (mass, NMR, and infrared spectra). The last of these compounds is probably not a direct ozonation product (stoichiometry and trapping studies). A mechanism is proposed in which the initial step is formation of a primary ozonide. From that stage, the breakdown is abnormal.

The structure of Feist's acid, Ia, was determined with certainty some 60 years after its preparation (*1–6*). More recently, the proposal that the ozonation product of Feist's ester (Ib) had structure II was shown to be incorrect (*7*). Our initial report not only showed that no II was present in any ozonation of Feist's ester but offered chemical and spectroscopic proof for III as a major ozonation product.

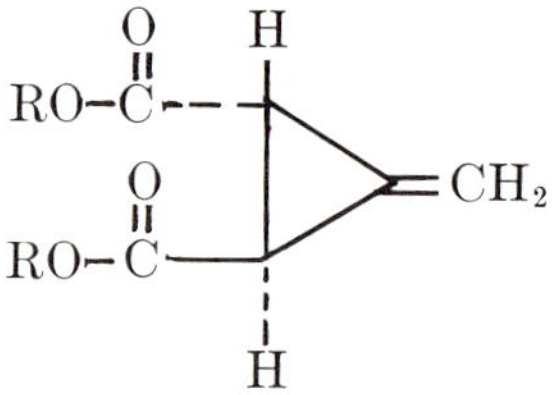

Ia, R=H
Ib, R=CH_3

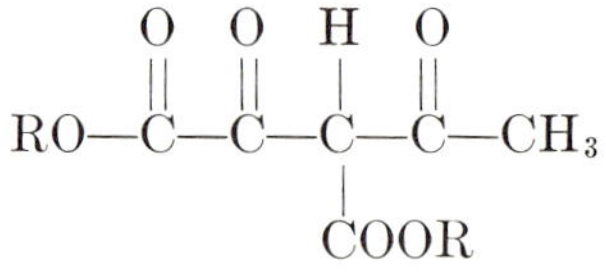

II

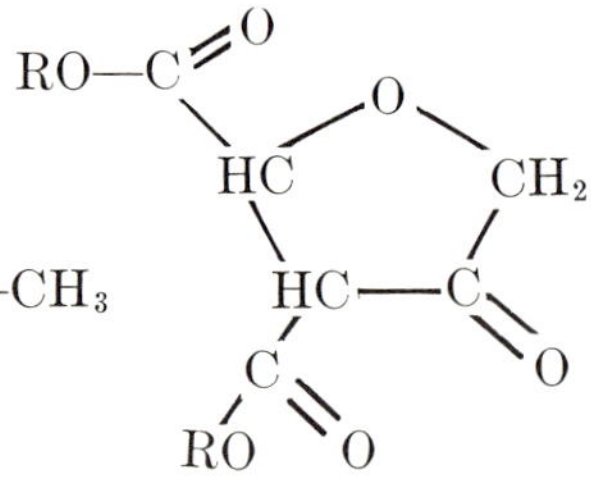

III

This paper is concerned with the structure proof of each of the major products of the ozonation of the methyl ester of Feist's acid. Other data pertaining to the mechanism of this unusual ozonation are presented, and a mechanism is hypothesized.

Experimental

Materials. Solvents were the best commercial grades available and were not purified further. Feist's acid was prepared by the method of Goss *et al.* (*2*). The methyl ester was prepared in the normal Fisher manner with methanol and sulfuric acid. Addition of water, ether extraction, removal of ether, and distillation yielded a colorless liquid (bp 78°–80°C at 1.5 mm Hg) which crystallized upon standing, mp 30°C (lit. value 30°C) (*8*).

2,3-DICARBOMETHOXYMETHYLENECYCLOPROPANE-2,3*d*2. The methyl ester of Feist's acid was specifically deuterated by dissolving 5.0 grams of the ester in 25 ml of deuteromethanol containing 0.5 gram sodium methoxide and allowing the mixture to stand for one day. After adding 40 ml of dilute sulfuric acid and extracting with two 20-ml portions of methylene chloride, a distillation yielded 3.5 grams of the dideutero compound. Its NMR spectrum showed only the methoxy protons (singlet at $\delta = 3.63$) and methylene protons (singlet at $\delta = 5.53$).

Ozonations. The ozonations were carried out using a Welsbach T-23 ozonator at −78°C in methanol, methylene chloride, Freon 11, or acetone. No changes in product ratios or ozone stoichiometry were noted when nitrogen was used as the ozone carrier.

Isolation of Products. Product isolation proved to be extremely difficult. A typical procedure involved the ozonation of the ester (2.0 grams) in 30 ml methanol at −78°C until the solution became blue and a precipitate had formed. After flushing with nitrogen, and allowing the mixture to stand overnight at −78°C, we collected 1.0 gram of precipitate. This was recrystallized three times from 15 ml of anhydrous ether to yield III, mp 67°–71°C (lit. 68°C). Evaporation of the first ether recrystallization residue and slow crystallization from 1.5 ml of anhydrous ether yielded compound IV, mp 96°–102°C [calculated for $C_8H_{10}O_7$: C = 44.04, H = 4.58, O = 51.8; found: C = 44.67, H = 4.66]. Compound V was isolated by ozonation of the ester in Freon 11 at −78°C and allowing the precipitate which formed to stand in solution overnight. After the precipitate was collected in a cold Buchner funnel (the precipitate melts below 0°C and was shown to be a mixture of III and IV by NMR), the filtrate was evaporated and the liquid residue was distilled to yield (bp 95°–100°C, 1 mm) a mixture rich in V (>90% according to NMR) [calculated for $C_8H_{10}O_5$: C = 51.63, H = 5.34, O = 43.03; found: C = 50.24, 50.53, H = 5.27, 5.29].

Product Analysis. Quantitative analysis for each of the products of the ozonation of Feist's ester was carried out by an NMR method (Varian HA60). The solvent for the ozonation was evaporated *in vacuo*, a suitable internal standard (usually *tert*-butyl benzoate), was added and the spectra were determined and integrated in chloroform-*d*. All analyses were reproducible to ±10%.

Table I. Important Ions in the Mass Spectra of III and IIIa

III		IIIa	
Assignment	m/e (*% rel. abundance*)	*Assignment*	m/e (*% rel. abundance*)
P	202 (2)	P	204 (1)
P-H_2O	184 (3)	P-HDO	185 (3)
P-OCH_3	171 (9)	P-OCH_3	173 (4)
P-CH_3OH	170 (53)	P-CH_3OD	171 (60)
P-$C_2H_3O_2$ and $\{OCH_2CO\}H$	143 (73)	P-$C_2H_3O_2$	145 (21)
		P-$\{OCH_2CO\}D$	144 (75)
P-$C_2H_3O_2$, CH_2O, and $\{C_2H_3O_2CHO\}H$	113 (100)	P-$C_2H_3O_2$, CH_2O	115 (33)
		P-$\{C_2H_3O_2CDO\}H$	114 (100)
P-$C_2H_3O_2$, CH_3OH	111 (53)	P-$C_2H_3O_2$, CH_2OD	112 (63)

Results

The structure of III was reasonably well established in our preliminary communication of this work. A comparison of its mass spectrum (Table I) with that of its dideutero analog IIIa confirms its structure and provides a mass spectral model for comparison with unknown compounds IV and V.

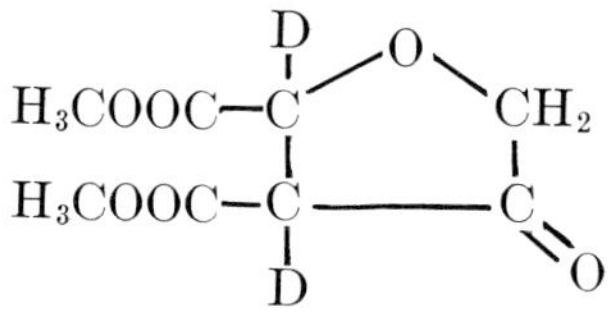

IIIa

The high mass peaks in Table I can be rationalized quite easily. However, the ions of highest abundance at 143 and 113 in III which at first glance can be assigned to fragmentation of a carbomethoxy group are, from inspection of the data from IIIa, of more complex origin. Compound IIIa had a peak at 145 as expected for loss of carbomethoxy, but it also shows an intense peak at 144 which must come from the loss of $\{—OCH_2C{=}O\}$ D. The deuterium theoretically could be transferred from either the α or β carbons to either the keto or ether functions. None of these rearrangements are particularly facile according to the literature. The base peaks in the spectra of III and IIIa (113 and 114 respectively) are also of interest. Loss of carbomethoxy and CH_2O, which appears to be a likely fragmentation pattern, does not seem to occur to a major extent for IIIa where the corresponding 115 peak is present in only 33% relative abundance. Thus, another hydrogen transfer, this time from the methylene group to one of the oxygen atoms, must take place.

The infrared spectrum of IV indicates an enol (3200 and 1650 cm^{-1}), an α,β unsaturated ester (1675 cm^{-1}), and a normal carbomethoxy carbonyl group (1725 cm^{-1}). The NMR spectrum substantiates the enolic character of the molecule with a singlet ($\delta = 11.3$ ppm). It also shows two doublets ($\sim$4.40 and 4.84 ppm, $J = 16.5$ cps) which can be assigned to two methylene protons (similar chemical shift to analogous protons in III), and singlets at 3.78 ppm and 5.16 ppm. The ratio of protons was shown to be 1:1:1:6:1 in the order given. Such a spectrum is consistent with either the enol form of III (IIIb shown below) or with structure IV.

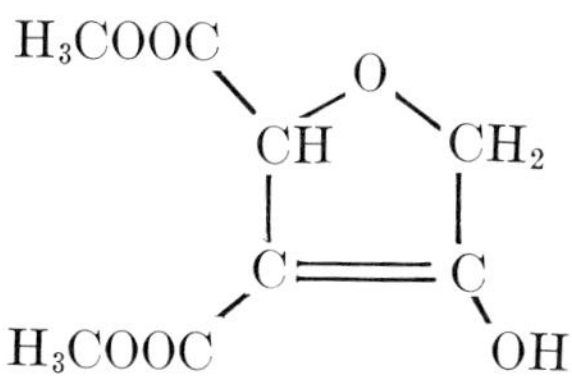

IIIb

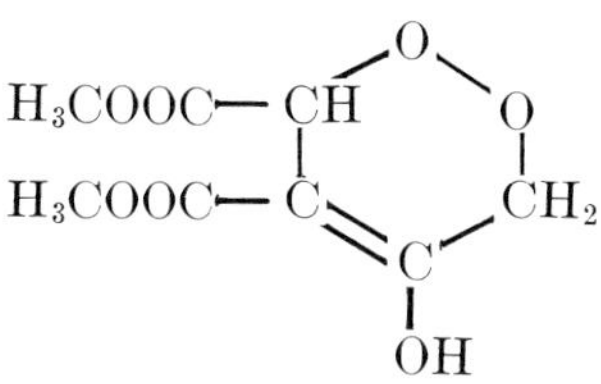

IV

Compound IV gave a positive peroxide test with potassium iodide in methanol and analyzed reasonably well for the structure shown. However, it was difficult to purify and decomposed upon standing, and we believed that mass spectral evidence was necessary for its final structure proof. We experienced some difficulty in obtaining pure dideutero IV but were able to obtain a satisfactory monodeutero compound by exchange of the enolic pattern with methanol-*d*. Table II compares the mass spectrum of IV with that of the monodeutero derivative.

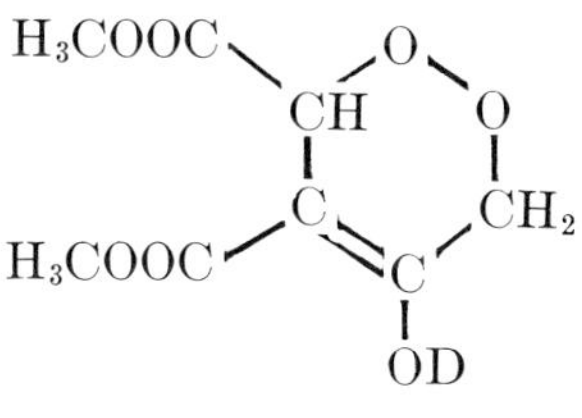

IVa

Although the parent, parent–OH(D), and parent–H_2O (HDO) peaks give strong corroborative evidence for IV and IVa as the correct

Table II. Important Ions in the Mass Spectra of IV and IVa

IV		*IVa*	
Assignment	m/e (% *rel. abundance*)	*Assignment*	m/e (% *rel. abundance*)
P	218 (4)	P	219 (4)
P-OH	201 (1)	P-OD	201 (2)
P-H_2O	200 (3)	P-HDO	200 (3)
P-OCH_3	187 (1)	P-OCH_3	188 (1)
P-$COOCH_3$	159 (100)	P-$COOCH_3$	160 (100)
P-$C_3H_7O_3$ or $C_2H_3O_4$	127 (44)	P-$C_3H_7O_3$ P-$C_2H_3O_4$	128 (17)
		P-$C_3H_6DO_3$ or $C_2H_2DO_4$	127 (25)

structure for the high melting compound, interpretation of the complete mass spectrum is not straightforward. For example, the base peak for both IV and IVa is accounted for readily by a loss of carbomethoxy, but the second most intense *m/e*, 127, must arise from at least two different pathways (neither of which is obvious). The accumulated evidence leaves little doubt but that IV is the correct structure of the high melting product.

The third compound to be isolated from ozonations of Feist's ester was V. Its NMR spectrum was of prime importance in establishing its structure. Other than the methoxy signal at 3.67 ppm, the outstanding feature of the spectrum was the presence of two sets of AB doublets with chemical shifts for the four protons at 3.33, 3.23, 2.90, and 2.72 ppm. The ratio of peak intensities was 6:1:1:1:1 for the five types of protons. In the spectrum of the dideutero derivative, the AB pattern at high field was not present, showing that these high field protons were those α to the carbomethoxy groups.

Table III. Important Ions in the Mass Spectra of V and Va

V		*Va*	
Assignment	m/e (% *rel. abundance*)	*Assignment*	m/e (% *rel. abundance*)
P	186 (0)	P	188 (0)
P-CH_2O	156 (2)	P-CH_2O	158 (2)
P-OCH_3	155 (11)	P-OCH_3	157 (12)
P-CH_3OH	154 (12)	P-CH_3OD	155 (13)
P-$C_2H_3O_2$	127 (100)	P-$C_2H_3O_2$	129 (100)
P-$C_2H_3O_2$, CH_2	113 (40)	P-$C_2H_3O_2$, CH_2	115 (44)

The major features of the mass spectra of V and its dideutero derivative are presented in Table III. Although we were unable to detect the parent peak for V, the reasonable assignments in Table II, the simplicity of the NMR spectrum, a comparison with normal epoxide chemical shifts (methylene signals at lower field than methoxy signals), and the analytical data argue for 4,5-dicarbomethoxy-1-oxaspiro[2.2]pentane as compound V.

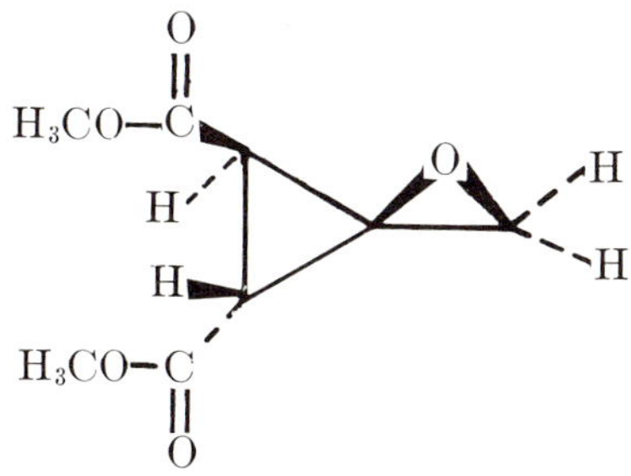

V

Table IV summarizes some of the spectral data for the assignment of structure to compounds III, IV, and V.

Table V shows the relative amounts of III, IV, and V found in different solvents under variable experimental conditions.

Knowledge of the major products of the ozonation of Feist's ester does not lead immediately to a knowledge of the mechanism of the reaction although the nature of the products makes it obvious that the mechanism is different from that of most ozonations. However, inspection of Table IV, in which product ratios change with changes in conditions and other types of experiments, allows us to rationalize a reasonable mechanistic sequence. The most important points to note are:

(1) The epoxide is not a primary product; both the stoichiometry and the tetracyanoethylene experiments indicate that it is formed from attack of an intermediate on unreacted Feist's ester. For example, less than 1 mole of ozone per mole of Feist's ester is required for all ozonations—with the less reactive solvents, such as methylene chloride and Freon 11, the deficiency in ozone required is approximately equal to the amount of epoxide formed. Addition of tetracyanoethylene before ozonation results in an ozonation of normal stoichiometry with no epoxide being formed.

(2) If the reaction is stopped before completion, the relative amounts of IV formed are greatly decreased. This suggests to us that the initial ozonation product has several modes of decomposition available, and that reaction with Feist's ester (or tetracyanoethylene) is more rapid than unimolecular decomposition.

(3) We were unable to detect an intermediate [*e.g.*, Criegee's primary ozonide (9)] by low temperature NMR (to −130°C); yet the above-mentioned results and the fact that IV does not react with Feist's ester strongly suggest that a highly reactive intermediate does exist.

Table IV. Spectral

Compound	*C–H*	*Mp, °C*	*MW*
III	$C_8H_{10}O_6$	67–71	202
IV	$C_8H_{10}O_7$	96–102	218
V	$C_8H_{10}O_5$	liquid	186

[a] As reported in Ref. *7*.

We hypothesize the following general reaction scheme:

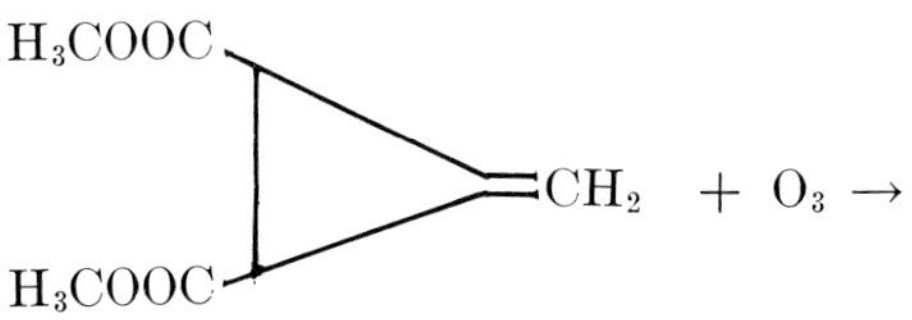

Ia

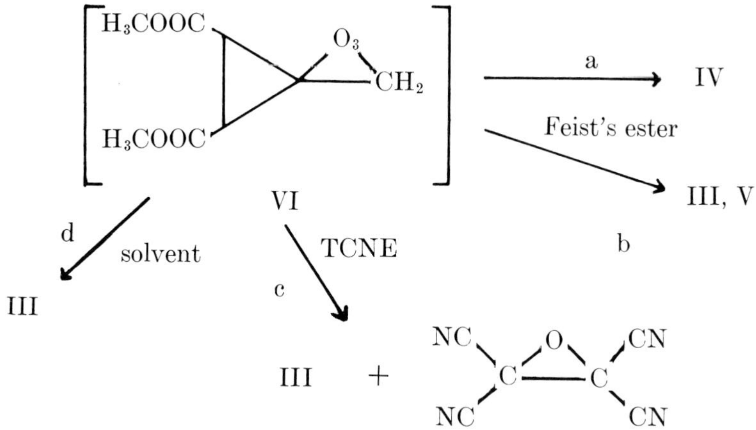

Path a involves unimolecular decomposition of an initial ozonide (VI). It is important to note that since other methylenecyclopropane derivatives, such as 2,3-dimethylmethylenecyclopropane, do not undergo ring opening of the cyclopropane ring, both ring strain and the presence of the carbomethoxy groups must be necessary for such an unusual decomposition of an initial ozonide. The following sequence rationalizes

Data for III, IV, and V

Infrared, cm^{-1}	*NMR* σ *mult. (rel. intensity)*
1778, 1740, 1675, 1636	5.1*d*(1), 4.25*d*(1), 4.12*d*(1),[a] 3.8*s*(6), 3.7*d*(1)
3200, 1725, 1675, 1600	11.3*s*(1), 5.16*s*(1), 4.84*d*(1), 4.55*d*(1), 3.78*s*(6)
3080, 3030, 1720, 890, 865, 833	3.67*s*(6), 3.33*d*(1), 3.23*d*(1), 2.90*d*(1), 2.72*d*(1)

such an occurrence by suggesting that a partial negative charge is stabilized by the carbomethoxy groups.

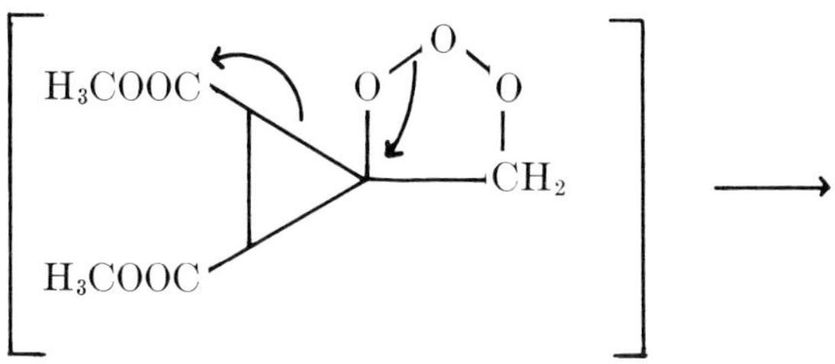

VI

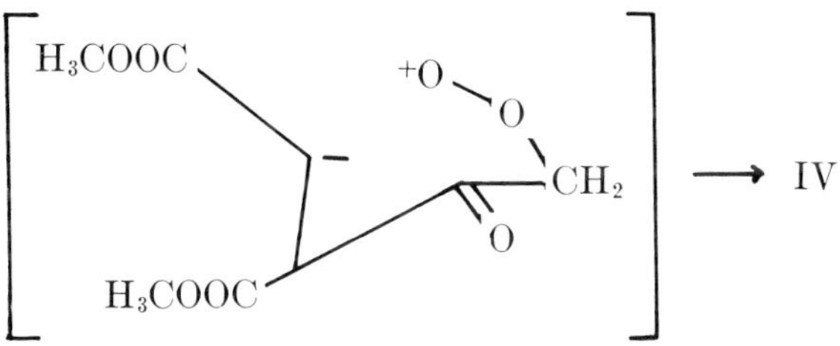

VII

Just as the conversion of VI to IV may be a concerted reaction without the intermediary action of VII, so pathway b might reasonably involve either VI or VII. Criegee and Günther have shown that tetracyanoethylene can change the normal course of the ozonation of alkenes by reaction with either a zwitterion or the "primary" ozonide produced in such reactions (*10*). We have no proof as to whether tetracyanoethylene reacts with VI or VII, and either reaction can be envisaged to produce III as in pathway c. We are left with the difficulty of explaining the relatively high yields of III in some solvents listed in Table IV. When III, a two-oxygen product is formed, either oxygen or some other oxidized

Table V. Relative Yields

Solvent	*Stoichiometry, moles O_3/moles I*	*Relative Yields, %* *III*	*IV*	*V*
CH_2Cl_2	0.67	40	21	39
CH_3OH	0.76	64	23	13
CH_3OH [a]	0.86	57	13	30
Freon 11	0.74	35	33	30
Freon 11 [b]	0.71	62	0	37
Acetone	0.75	38	30	32
Acetone–CH_2Cl_2	0.65	33	32	35
Acetone–CH_2Cl_2 [c, d]	1.03	71	29	0

[a] Reaction stopped after 60% of I had reacted.
[b] Reaction stopped after 34% of I had reacted.
[c] 75% CH_2Cl_2 by volume.
[d] Tetracyanoethylene (equimolar with I) added before reaction began.

product must be produced. With oxidizable solvents, such as methanol or acetone, the most reasonable assumption might be that VI or VII reacts with solvent to yield III (path d), but we have no direct proof for such an assumption.

Story *et al.* have suggested that VIII might be the important intermediate in the ozonation of Feist's ester (*11*).

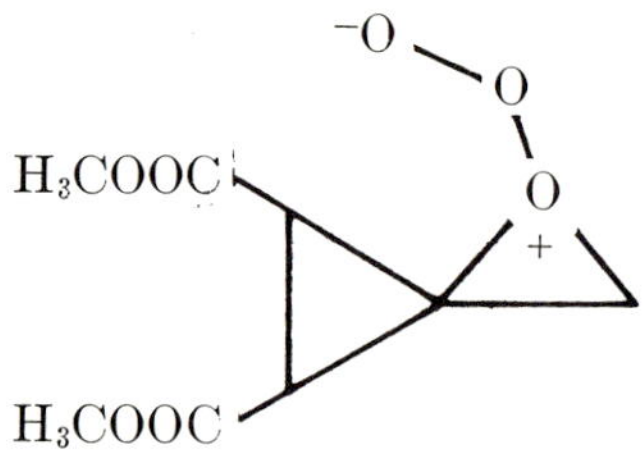

VIII

If this is the intermediate, other methylenecyclopropanes should give ozonation products similar to those found in the ozonation of Feist's ester. Since they do not and because our evidence indicates that the epoxide is not a primary product, the mechanism of Story *et al.* does not seem to apply to this reaction.

Literature Cited

1. Feist, F., *Chem. Ber.* (1893) **26,** 747.
2. Goss, F. R., Ingold, C. T., Thorpe, J. F., *J. Chem. Soc.* (1923) **123,** 327.

3. Ettlinger, M. G., *J. Amer. Chem. Soc.* (1952) **74**, 5805.
4. Ettlinger, M. G., Kennedy, F., *Chim. Ind.* (1956) 166.
5. Kende, A. S., *Chim. Ind.* (1956) 437.
6. Bottini, A. T., Roberts, J. D., *J. Org. Chem.* (1956) **21**, 1169.
7. Erickson, R. E., *Tetrahedron Letters* (1966) 1753.
8. Schwan, U., Ph.D. thesis, Technische Hochschule, Karlsruhe (1958).
9. Criegee, R., *Rec. Chem. Progr.* (1957) **18**, 111.
10. Criegee, R., Günther, P., *Chem. Ber.* (1963) **96**, 1564.
11. Story, P. R., Alford, J. A., Ray, W. L., Burgess, J. R., *Preprints, Div. Petroleum Chemistry, ACS* (1971) **16**, A13.

RECEIVED May 20, 1971. Supported by the Petroleum Research Fund, administered by the American Chemical Society.

INDEX

INDEX